T0141838

Cognitive Systems Monographs

Volume 30

Series editors

Rüdiger Dillmann, University of Karlsruhe, Karlsruhe, Germany
e-mail: ruediger.dillmann@kit.edu

Yoshihiko Nakamura, Tokyo University, Tokyo, Japan
e-mail: nakamura@ynl.t.u-tokyo.ac.jp

Stefan Schaal, University of Southern California, Los Angeles, USA
e-mail: sschaal@usc.edu

David Vernon, University of Skövde, Skövde, Sweden
e-mail: david@vernon.eu

About this Series

The Cognitive Systems Monographs (COSMOS) publish new developments and advances in the fields of cognitive systems research, rapidly and informally but with a high quality. The intent is to bridge cognitive brain science and biology with engineering disciplines. It covers all the technical contents, applications, and multidisciplinary aspects of cognitive systems, such as Bionics, System Analysis, System Modelling, System Design, Human Motion, Understanding, Human Activity Understanding, Man-Machine Interaction, Smart and Cognitive Environments, Human and Computer Vision, Neuroinformatics, Humanoids, Biologically motivated systems and artefacts Autonomous Systems, Linguistics, Sports Engineering, Computational Intelligence, Biosignal Processing, or Cognitive Materials as well as the methodologies behind them. Within the scope of the series are monographs, lecture notes, selected contributions from specialized conferences and workshops.

More information about this series at http://www.springer.com/series/8354

Boris Schauerte

Multimodal Computational Attention for Scene Understanding and Robotics

 Springer

Boris Schauerte
Karlsruhe
Germany

ISSN 1867-4925   ISSN 1867-4933   (electronic)
Cognitive Systems Monographs
ISBN 978-3-319-81605-0   ISBN 978-3-319-33796-8   (eBook)
DOI 10.1007/978-3-319-33796-8

Printed on acid-free paper

This Springer imprint is published by Springer Nature
The registered company is Springer International Publishing AG Switzerland

Foreword

While it might seem that humans are at each moment aware of everything they perceive through their senses, this is in fact not the case. Instead only a fraction of the sensory input that becomes aware to us is consciously experienced and is further attended to. The reason for this selective attention is assumed to be the limited processing capacity of our brains: We simply cannot consciously process all the sensory input at any given time. Researchers have investigated this phenomenon for more than a hundred years, and have come up with various models for the underlying mechanisms.

When we aim at building technical systems with visual or acoustic sensory processing capabilities, such as for example robots, we are faced with similar resource allocation problems. Due to limited computational processing capabilities, it is not possible to constantly process all sensory input to largest extent. Thus, computational attention models are needed to focus limited processing power on the most interesting parts of the sensory input first.

In this book, Boris Schauerte gives a comprehensive overview of state-of-the art computational attention models for processing visual and acoustic input. The book covers biological background of visual and auditory attention, as well as bottom-up and top-down attentional mechanisms. Also different applications areas are discussed. The book integrates more than five years of the author's research in this field and summarizes a dozen of his scientific publications on this topic. In this respect, the book contains a number of scientific contributions to the field: In the first part of the book, new approaches for bottom-up visual and acoustic saliency models are presented, and applied to the task of audio-visual scene exploration of a robot. In the second part, the influence of top-down cues for attention modeling is investigated. For example, how gaze and gestures can be used as top-down cues for modeling attention in a human–robot interaction scenario, or how the gaze of people depicted in pictures influences our own gaze behavior when looking at those pictures.

I am sure the book will be of great value to any reader who is interested in computational attention models. The book gives an excellent introduction to the field, it extensively discusses state-of-the art attentional models and it summarizes a number of the author's contributions to this exciting field.

Karlsruhe   Prof. Dr.-Ing. Rainer Stiefelhagen
January 2016

Contents

About the Author

Dr. Boris Schauerte received his PhD degree from the Karlsruhe Institute of Technology in 2014. As an entrepreneur and researcher he explored, applied, and made fundamental contributions to all aspects of computational attention using advanced statistics, signal processing and machine learning methods including computer vision, and deep learning. He has used his knowledge to enhance user experience analysis, customer insights, robotic perception, and smart assistive systems in research as well as in commercial projects.

Abbreviations

2D	Two-dimensional
3D	Three-dimensional
4D	Four-dimensional
AUC	Area under the curve
AUROC	Area under the receiver operator characteristic curve
CC	Correlation coefficient
CJAD	Cumulated joint angle distance
CRF	Conditional random field
CS	Cumulated saliency
DCT	Discrete cosine transform
EMD	Earth mover's distance
EPQFT	Eigen pure quaternion Fourier transform
ESR	Eigen spectral residual
ESW	Eigen spectral whitening
FFT	Fast Fourier transform
FHR	Focus of attention hit rate
FIT	Feature integration theory
FoA	Focus of attention
GBVS	Graph-based visual saliency
GPU	Graphics processing unit
GSM	Guided search model
HOF	Histogram of optical flow
HOG	Histogram of oriented gradients
HRI	Human–robot interaction
ICOPP	Intensity and color opponents
ICS	Integrated cumulated saliency
ICU	Intensive care unit
iNVT	iLab Neuromorphic Vision Toolkit
IoR	Inhibition of return
KIT	Karlsruhe Institute of Technology
KLD	Kullback–Leibler divergence

LBP	Local binary patterns
MCT	Modified census transform
MDCT	Modified discrete cosine transform
MRD	Medical recording device
MSER	Maximally stable extremal regions
nAUROC	Normalized area under the receiver operator characteristic curve
NCJAD	Normalized cumulated joint angle distance
NCS	Normalized cumulated saliency
NP	Nondeterministic polynomial time
NSS	Normalized scanpath saliency
NTOS	Normalized target object saliency
PCA	Principal component analysis
PGC	Probabilistic gaze cone
PHAT	Phase transform
PHR	Pixel hit rate
POS	Part-of-speech
PPC	Probabilistic pointing cone
PPCC	Probability plot correlation coefficient
PQFT	Pure quaternion Fourier transform
PTU	Pan-tilt-unit
QDCT	Quaternion discrete cosine transform
Q-Q	Quantile–quantile
RAM	Random access memory
ROC	Receiver operating characteristic
SNR	Signal-to-noise ratio
SRP	Steered response power
STCT	Short-time cosine transform
STFT	Short-time Fourier transform
STIP	Space time interest points
SVM	Support vector machine
TDOA	Time difference of arrival
TRW	Tree-reweighted belief propagation
TSP	Traveling salesman problem
VOCUS	Visual object detection with computational attention system
WTA	Winner-take-all
ZCA	Zero-phase transform

List of Figures

List of Tables

Abstract

Attention is the cognitive process that identifies subsets within sensory inputs (e.g., from the millions of human sensory receptors) that contain important information to focus subsequent complex and slow processing operations on the most relevant information. This is a key capability of humans and animals that allows us to rapidly understand what is going on in a scene despite the limited computational capacities of the brain. Consequently, since attention serves as a gateway to later cognitive processes, efficient, reliable, and rapid attentional allocation is key to predation, escape, and mating—in short, to survival.

Like their biological counterparts, robotic systems have limited computational capacities. Consequently, computational attention models are important to allow for complex cognitive processing. For this purpose, we develop highly efficient auditory and visual attention models. For visual attention, we use hypercomplex image processing and decorrelation to calculate what is interesting in an image and are able to efficiently predict where people will look in an image. For auditory attention, we use Bayesian methods to determine what are unexpected and thus surprising sounds. Here, we are able to reliably detect arbitrary, salient acoustic events. We fuse the auditory and visual saliency in a crossmodal parametric proto-object model. Based on the detected salient proto-objects in a scene, we can use multiple criteria to plan which part of the room the robot should attend next. We have successfully implemented this approach on robotic platforms to efficiently explore and analyze scenes.

In many situations, people want to guide our attention to specific aspects. For example, photographers compose their images in such way that the most important object automatically grabs the viewer's attention. Furthermore, people use non-verbal signals (e.g., pointing gestures and gaze) to control where a conversation partner looks to include a specific nearby object in the conversation. Interestingly, infants develop the ability to interpret such non-verbal signals very early and it is an essential ability, because it allows to associate spoken words with the visual appearances of nearby objects and thus to learn language.

In the second part, we first try to identify the most prominent objects in web images. Then, we start to integrate verbal and non-verbal social signals into our saliency model. As non-verbal signals, we consider gaze and pointing gestures. Both signals direct our attention toward spatial areas to narrow the referential domain, which we model with a probabilistic corridor of attention. As verbal signals, we focus on specific spoken object descriptions that have been shown to being able to directly guide visual saliency and thus influence human gaze patterns. Interestingly, verbal and non-verbal signals complement each other, i.e., as one signal type becomes ambiguous it is compensated with the other. We achieve the best results with machine learning methods to integrate the available information. This way, we are able to efficiently highlight the intended target objects in human–robot interaction and web images.

Chapter 1
Introduction

We immediately spot a warning triangle on a street or a black sheep in a flock. Yet, although we know what we are looking for, it can take us minutes to find Waldo, who blends into a crowd of nondescript people. When it comes to hearing, we are able to selectively listen to different speakers in a crowded room that is filled with a multitude of ongoing conversations. And, an unexpected, unfamiliar sound at night can awaken and scare us, making our hearts race as a means to prepare us for fight or flight. These examples illustrate how our brain highlights some visual or auditory signals while suppressing others. Understanding what our brain will highlight is not just fundamental to understand and model the human brain but forms the basis for best practices in various application areas. For example, based on a set of basic cognitive rules and guidelines, movie directors compose the camera shots of scenes in such a way that the relevant information gets subconsciously highlighted. Furthermore, horror movies use harsh, non-linear, and unexpected sounds to trigger strong emotional responses.

This form of highlighting is better known as "selective attention" and describes mechanisms in the human brain that determine which parts of the incoming sensory signal streams are currently the most interesting and should be analyzed in detail. Attentional mechanisms select stimuli, memories, or thoughts that are behaviorally relevant among the many others that are behaviorally irrelevant. Such attentional mechanisms are an evolutionary response to the problem that the human brain—due to computational limitations—is not able to fully process all incoming sensory information and, as a consequence, has to select and focus on the potentially most relevant stimuli. Otherwise humans would not be able to rapidly understand what is going on in a scene, which however is key to predation and escape—in short, to human survival and evolution.

Thus, attention serves as a gateway to later cognitive processes (e.g., object recognition) and visual attention is often compared with a "spotlight". Following the spotlight metaphor, only scene elements that are illuminated by the spotlight are fully processed and analyzed. By moving the spotlight around the scene, we can iteratively build up an impression of the entire scene. For example, in the human visual

© Springer International Publishing Switzerland 2016
B. Schauerte, *Multimodal Computational Attention for Scene Understanding and Robotics*, Cognitive Systems Monographs 30,
DOI 10.1007/978-3-319-33796-8_1

system this is implemented in the form of rapid, subconscious eye movements, the so-called "saccades". By moving the eye, fixating and analyzing one location at a time, the small fixated part of the scene is projected onto the fovea. The fovea is the central part of the retina that is responsible for highly resoluted, sharp, non-peripheral vision. As a consequence, attention does not just reduce the necessary computational resources, but it ensures the best possible sensory quality of the fixated sensory stimuli for subsequent stages—thus, it represents an evolutionary solution to manage perceptual sensory quality and computational limitations. Orienting the eyes, head, or even body to selectively attend a stimulus is called "overt attention". In contrast, "covert attention" describes a mental focus (e.g., to focus on a specific aspect of an overtly focused object) that is not accompanied by physical movements.

Since only a small part of the signal will be analyzed, the definition of what is potentially relevant—i.e., "salient"—is absolutely critical. Here, we have to differentiate between two mechanisms. First, bottom-up, stimulus-driven saliency highlights signals as being salient that differ sufficiently from their surrounding in space and time. For example, due to its unnatural triangular shape and color, the advance warning triangle is highly salient; as is the black sheep that visually "pops out" of the flock of white sheep. Similarly, a sudden, unexpected sound attracts our auditory attention, because it differs substantially from what we have heard before. Bottom-up attention is also often referred to as being "automatic", "reflexive", or "stimulus-driven". Second, top-down, user-driven factors can strongly modulate or, in some situations, even override bottom-up attention. Such top-down factors can be expectations or knowledge about the appearance of a target object that is being searched (i.e., the basis for so-called "visual search") that influences which distinctive features should attract our attention. For example, during a cocktail party we are able to focus our auditory attention on a location (i.e., the location of the person we want to listen to) and specific frequencies to highlight the voice of our conversation partners and suppress background noise, which allows us to better understand what is being said. Similarly, for example, when visually searching for a red object, all red objects in the scene become more salient. However, top-down attention also faces limitations that can be experienced when looking for Waldo, which—due to the presence of distractors—is still a challenging problem even though we exactly know how Waldo looks like. Furthermore, in many situations, bottom-up attention can not be suppressed entirely and highly salient stimuli can still attract the attention independent of conflicting top-down influences. Top-down attention is also commonly referred to as being "voluntary", "centrally cued", or "goal-driven".

Naturally, auditory and visually salient stimuli are integrated into a crossmodal attention model and work together. For example, when we hear a strange, unexpected sound behind our back, then we will naturally turn our head to investigate what has caused this sound. Furthermore, information that we acquire from speech (e.g., about the visual appearance of an object) can modulate the visual saliency. In fact, in recent years, it becomes more and more apparent that the sensory processing in the human brain is multisensory to such extent that, for example, lipreading or the observation of piano playing without hearing the sound can activate areas in the auditory cortex.

Attention models try to model what the human brain considers as being salient or interesting. Traditionally, attention models have been used to model and predict the outcome of psychological experiments or tests with the goal to understand the underlying mechanisms in the human brain. However, attention models are not just interesting to achieve a better understanding of the human brain, because to know what humans find interesting is an important information for a wide range of practical applications. For example, we could optimize the visual layout of advertisement or user interfaces, reduce disturbing signal compression artifacts, or suppress annoying sounds in urban soundscapes. In general, knowing what is potentially relevant or important information opens further application scenarios. For example, we could focus machine learning algorithms on the most relevant training data. An application area that seems to be particularly in need of attention mechanisms is robotics, because robots that imitate aspects of human sensing and behavior face similar challenges as humans. Accordingly, attention models could be used to implement overt and covert attention to save computational resources, improve visual localization, or help to mimic aspects of human behavior in human-robot interaction.

In this book, we describe our work on two aspects of multimodal attention:

1. Bottom-up Audio-Visual Attention for Scene Exploration
    In the first part, we describe how we realized audio-visual overt attention on a humanoid robot head. First, we define which auditory and visual stimuli are salient. For this purpose, we use spectral visual saliency detection with a decor-related color space for visual saliency and a probabilistic definition of surprise to implement auditory saliency detection. Then, we determine and localize auditory and/or visually salient stimuli in the robot's environment and, for each salient stimulus, we represent the spatial location and extent as well as its saliency in the form of so-called proto-objects. This makes it possible to fuse the auditory and visual proto-objects to derive crossmodally salient regions in the environment. Based on these salient spatial regions, i.e. our salient proto-objects, we implement overt attention and plan where the robot should turn its head and look next. Here, we do not just incorporate each proto-object's saliency, but use a multiobjective framework that allows us to integrate ego-motion as a criterion.

2. Multimodal Attention with Top-Down Guidance
    In the second part, we investigate attention models for situations in which a person tries to direct the attention toward a specific object, i.e. an intended target object. Here, we address two top-down influences and application domains: First, how photographers and other artists compose images to direct the viewer's attention toward a specific salient object that forms the picture's intended center of interest. Second, how interacting people use verbal (e.g., "that red cup") and non-verbal (e.g., pointing gestures) signals to direct the interaction partner's visual attention toward an object in the environment to introduce or focus this object in the conversation and talk about it. For this purpose, we rely on machine learning methods to integrate the available information, where we also build on features that we derived in the first part of this book. Finally, we combine both tasks and address web images in which people are looking at things. Thus, we shift from the first

part's focus on the general interestingness of signals to being able to let top-down information guide visual saliency and highlight the specific image regions that depict intended target objects.

1.1 Contributions

Among several other contributions, our major contributions to the state-of-the-art that we present and discuss in this book have been made in these areas:

1. Visual saliency
   We focused on how we can represent color information in a way that supports bottom-up visual saliency detection. For this purpose, we investigated the use of quaternions for holistic color processing, which in combination with quaternion component weighting was able to improve the state-of-the-art in predicting where people look by a small margin. Based on our experiences with quaternion-based approaches, we investigated color decorrelation as a method to represent color information in a way that supports to independently process color channels. This way, we improved the predictive performance of eight visual saliency algorithms, again improving the state-of-the-art.

2. Auditory saliency
   We proposed a novel auditory saliency model that is based on Bayesian surprise. Our model has a clear biological foundation and, in contrast to prior art, it is able to detect salient auditory events in real-time. The latter was an important requirement to implement auditory attention on a robotic platform. Since a similar approach has not been proposed and evaluated before, we also introduced a novel, application-oriented evaluation methodology and show that our approach is able to reliably detect arbitrary salient acoustic events.

3. Audio-visual proto-objects and exploration
   Proto-objects are volatile units of information that can be bound into a coherent and stable object when accessed by focused attention, where the spatial location serves as index that binds together various low-level features into proto-objects across space and time. We introduce Gaussian proto-objects as novel, object-centred method to represent the 3-dimensional spatial saliency distribution. In contrast to prior art, our representation is parametric and not grid-like such as, for example, elevation-azimuth maps or voxels. We implement proto-objects as being primitive, uncategorized object entities in our world model. This way proto-objects seamlessly form the foundation to realize biologically-plausible crossmodal saliency fusion, implement different overt attention strategies, realize object-based inhibition of return, and serve as starting point for the hierarchical, knowledge-driven object analysis. We are not aware of any prior system that integrates all these aspects in a similarly systematic, consistent, biologically-inspired way.

4. Salient object detection
   We investigated how the photographer bias influences salient object detection datasets and, as a consequence, algorithms. We provided the first empirical justification for the use of a Gaussian center-bias and have shown that algorithms may have implicit, undocumented biases that enable them to achieve better results on the most important datasets. Based on these observations, we adapted a state-of-the-art algorithm and removed its implicit center-bias. This way, we were able to achieve two goals: First, we could improve the state-of-the-art in salient object detection on web images through the integration of an explicit, well-modeled center-bias. Second, we derived the currently best performing unbiased algorithm, which can provide superior performance in application domains in which the image data is not subject to a center-bias.

5. Saliency with top-down guidance
   We were the first to create attention models that let the often complementary information contained in spoken descriptions of a target object's visual appearance and non-verbal signals—e.g., pointing gestures and gaze—guide the visual saliency and, as a consequence, the focus of attention. This way, we are often able to highlight the intended target object in human-robot interaction with the goal to facilitate to establish a joint focus of attention between interacting people. We started with biologically-oriented models, but achieved the best results with machine learning methods that learn how to integrate different features, ranging from our spectral visual saliency models to probabilistic color term models. After having demonstrated that this successfully works for human-robot interaction, we approach a more challenging domain and try to identify the objects of interest in web images that depict persons looking at things.

We provide a more detailed discussion of individual contributions in the related work part of Chaps. 3 and 4.

Code and Impact

To support scholarly evaluation by other researchers as well as the integration of our methods into other applications, we made most algorithms that are described in this book open source. This includes, for example, the source code for auditory saliency detection—including Gaussian surprise—and our spectral visual saliency toolbox. The latter was downloaded several thousand times during the past years. Our code has also been successfully used in other research projects at the computer vision for human-computer interaction lab:

1. Patient agitation
   Our Gaussian surprise model was used for patient agitation detection in intensive care unit, see Sect. A.1.
2. Activity recognition
   Our quaternion image signature saliency model and proto-objects were used to improve activity recognition, see Sect. A.2.

Further Contributions

Related to some of the content presented in this book, but not thematically central enough to be described in detail, we contributed to further fields:

1. Assistive technologies for visually impaired people

   Initially, we learned and applied color term models to integrate the top-down influence of spoken object descriptions on attention in human-robot interaction [SF10a]. However, these color models also became an essential element in our work on computer vision for blind people, because they allowed us to use sonification to guide a blind person's attention toward certain spatial areas and help find lost things [SMCS12].

   Furthermore, as part of this book, we use conditional random fields to learn to guide visual saliency in human-robot interaction, see Sect. 4.3. However, we also have applied this conditional random field structure, learning, and prediction methodology to identify the area in front of a walking person that is free of obstacles (see [KSS13]).

2. Color term and attribute learning from web images

   We learned color term models (i.e., representations of what people actually mean when they say "red" or "blue") from images that were automatically gathered from the web. Here, we proposed to use image randomization in such a way that the color distributions of artificial or post-processed images, which are common in the domain of web images, better match the distribution of natural images [SF10b]. To further improve the results, we combined salient object detection as spatial prior and supervised latent Dirichlet allocation to treat color terms as linguistic topics [SS12a]. Inspired by our intended target application, we introduced an evaluation measure that compares the classification results with human labels, which better reflects the oftentimes fuzzy boundaries between color terms.

1.2   Outline

The organization of this book is as follows:

Main Content

In Chap. 2, we provide a broad overview of related work that forms the background to understand many ideas and concepts throughout this book. We first discuss aspects of auditory (Sect. 2.1.1), visual (Sect. 2.1.2), and multimodal attention (Sect. 2.1.3). Then, we overview attention model applications (Sect. 2.2).

In Chap. 3, we present how we implemented audio-visual bottom-up attention for scene exploration and analysis. Here, we start with a discussion of the most relevant related work and how our approach deviates from prior art (Sect. 3.1). Then, we introduce how we define visual and auditory saliency (Sects. 3.2 and 3.3, respectively) before we explain how we fuse this information in an audio-visual proto-object model (Sect. 3.4). The proto-object model forms the basis to plan where the robot should look next, which is implemented solely saliency-driven (Sect. 3.4.4.A) and with the consideration of the necessary ego-motion (Sect. 3.5).

In Chap. 4, we present how we learn to let top-down influences guide attention to highlight specific objects of interest. Again, we first discuss the most relevant related work and clarify our contributions (Sect. 4.1). Then, we analyze how the photographer bias in web images influences modern salient object detection algorithms and present a state-of-the-art method without such a bias (Sect. 4.2). Afterwards, we discuss how we integrate pointing gestures and spoken object references into an attention model that highlights the referred-to object (Sect. 4.3). Finally, we show how the methods that we first presented in an human-robot interaction context (Sect. 4.3) can be applied to web images to identify objects that are being looked at.

In Chap. 5, we summarize the results presented in this book and discuss potential aspects of future work.

Appendices

In Appendix A, we present two applications that rely on our saliency models and were developed at the computer vision for human-computer interaction lab. First, in Sect. A.1, we describe how Martinez uses our Gaussian surprise model to detect patient agitation in intensive care units. Second, in Sect. A.2, we describe how Rybok uses our quaternion image signature saliency model and proto-objects to improve the accuracy of an activity recognition system.

In Appendix B, we overview and briefly describe all twelve datasets that are relevant to this book.

In Appendix C, we provide further color space decorrelation results that supplement our evaluation in Sect. 3.2.2.B.

References

[KSS13]   Koester, D., Schauerte, B., Stiefelhagen, R.: Accessible section detection for visual guidance. In: IEEE/NSF Workshop on Multimodal and Alternative Perception for Visually Impaired People (2013)
[SF10a]   Schauerte, B., Fink, G.A.: Focusing computational visual attention in multi-modal human-robot interaction. In: Proceedings of the 12th International Conference on Multimodal Interfaces and 7th Workshop on Machine Learning for Multimodal Interaction (ICMI-MLMI). ACM, Beijing, China, Nov 2010
[SF10b]   Schauerte, B., Fink, G.A.: Web-based learning of naturalized color models for human-machine interaction. In: Proceedings of the 12th International Conference on Digital Image Computing: Techniques and Applications (DICTA). IEEE, Sydney, Australia, Dec 2010
[SS12a]   Schauerte, B., Stiefelhagen, R.: Learning robust color name models from web images. In: Proceedings of the 21st International Conference on Pattern Recognition (ICPR). IEEE, Tsukuba, Japan, Nov 2012
[SMCS12] Schauerte, B., Martinez, M., Constantinescu, A., Stiefelhagen, R.: An assistive vision system for the blind that helps find lost things. In: Proceedings of the 13th International Conference on Computers Helping People with Special Needs (ICCHP). Springer, Linz, Austria, July 2012

Chapter 2
Background

Although in principle all attention models serve the same purpose, i.e. to highlight potentially relevant and thus interesting—that is to say "salient"—data, attention models can differ substantially in which parts of the signal they mark as being of interest. This is to a great extent due to the varying research questions and interests in relevant fields such as, most importantly, neuroscience, psychophysics, psychology, and computer science. However, it is also caused by the vagueness as well as application- and task-dependence of the underlying problem description, i.e. what is interesting?

The purpose of this chapter is to provide an introduction to visual and auditory attention (Sect. 2.1) and its applications (Sect. 2.2) that serves as background information for the remainder of this book.

2.1 Attention Models

In general, it is possible to distinguish three types of attention models by the respective research field: First, neurobiological models try to understand and model in which part of the brain attentional mechanisms reside and how they operate and interact on a neurobiological level. Second, psychological models try to model, explain, and better understand aspects of human perception and not the brain's neural system and layout. Third, computational models implement principles of neurobiological and psychological models, but they are also often subject to an engineering objective. Such an engineering objective is less to model the human brain or perception, but to be part of and improve artificial systems such as, e.g., vision systems or complex robots.

For visual attention, the following text focuses on computational and to a lesser extent psychological models, because well-studied, elaborated psychological and computational models exist. Furthermore, a deep understanding of neurobiological aspects of the human brain's neural visual system is of minor relevance for the

© Springer International Publishing Switzerland 2016
B. Schauerte, *Multimodal Computational Attention for Scene Understanding and Robotics*, Cognitive Systems Monographs 30,
DOI 10.1007/978-3-319-33796-8_2

remainder of this book. An interesting complementary lecture to this section is the excellent survey by Frintrop et al. [FRC10], which specifically tries to explain attention related concepts and ideas across the related fields of neurobiology, psychology, and computer science. For auditory attention, it is necessary to address neurobiological aspects of the human auditory system, because concise elaborated psychological and computational do not exist and a basic understanding of the human auditory system is important to understand the motivation of proposed computational models. Here, Fritz et al. and Hafter et al. provide very good neurobiological overviews of auditory attention [FEDS07, HSL07].

2.1.1 Visual Attention

Psychlogical Models

The objective of psychological attention models is to explain and better understand human perception, not to model the brain's neural structure. Among the psychological models, the feature integration theory (FIT) by Treisman et al. [TG80] and Wolfe et al.'s guided search model (GSM) [Wol94] are probably by far the most influential models. Aspects of both models are still present in modern models and both models have constantly been adapted to incorporate later research findings. A deeper discussion of psychological models can be found in the review by Bundesen and Habekost [BH05].

Treisman's feature integration theory [TG80], see Fig. 2.1, assumes that "different features are registered early, automatically, and in parallel across the visual fields, while objects are identified separately and only at a later stage, which requires focused attention" [TG80]. This simple description includes various aspects that are still fundamental for psychological and computational attention models. Conspicuities in a feature channel are represented in topological "conspicuity" or "feature maps". The information from the feature map is integrated in a "master map of location". A concept that is nowadays most widely known as "saliency map" [KU85]. This master map of location encodes "where" things are in an image, but not "what" they are, which reflects the "where" and "what" pathways in the human brain [FRC10]. Attention is serially focused on the highlighted locations in the master map and the image data around the attended location is passed as data to higher perception tasks such as, most importantly, object recognition to answer "what" is shown at that location.

Although Treisman's early model primarily focused on bottom-up perceptual saliency, Treisman also considered how attention is affected during visual search, i.e. when looking for specific target objects. A target is easier—i.e., faster—to find during visual search the more distinctive features it exhibits that differentiate it from the distractors. To implement visual search mechanism in FIT, Treisman proposed to inhibit the feature maps that encode the features of distractors, i.e. non-target features.

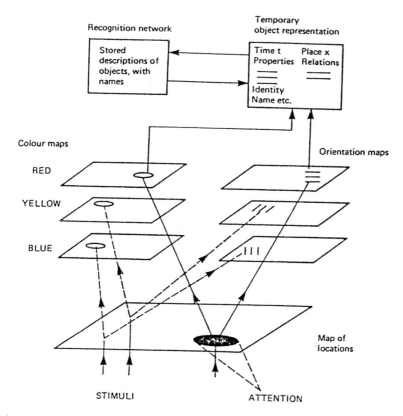

Fig. 2.1 Treisman's feature integration theory model. Image from [TG88] reprinted with permission from APA

Treisman et al. also introduced the concept of object files as "temporary episodic representations of objects". An object file "collects the sensory information that has so far been received about the object. This information can be matched to stored descriptions to identify or classify the object" [KTG92].

Wolfe et al. [WCF89, Wol94] introduced the initial guided search model to address shortcomings of early versions of Treisman's FIT model, see Fig. 2.2. As its name suggests, Wolfe's GSM focuses on modeling and predicting the results of visual search experiments. Accordingly, it explicitly integrates the influence of top-down information to highlight potential target objects during visual search. For this purpose, it uses the top-down information to select the feature type that best distinguishes between target and distractors.

Computational Models—Traditional Structure

Most computational attention models follow a similar structure, see Fig. 2.3, which is adopted from Treisman's feature integration theory [TG80] and Wolfe's guided search model [WCF89, Wol94] (see Figs. 2.1 and 2.2, respectively). The first com-

Basic Components of Guided Search

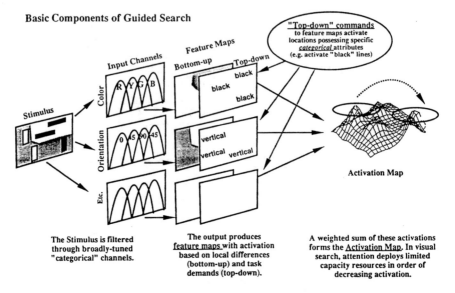

The Stimulus is filtered through broadly-tuned "categorical" channels.

The output produces feature maps with activation based on local differences (bottom-up) and task demands (top-down).

A weighted sum of these activations forms the Activation Map. In visual search, attention deploys limited capacity resources in order of decreasing activation.

Fig. 2.2 Wolfe's guided search model. Image from [Wol94] reprinted with permission from Springer

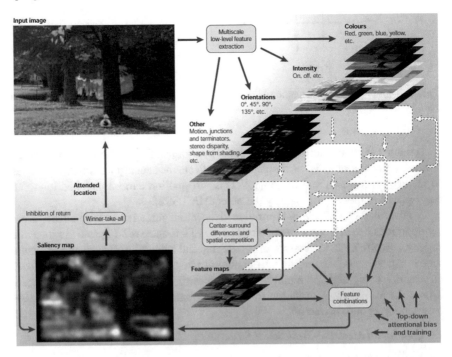

Fig. 2.3 The traditional structure of feature-based computational visual attention models, which is the basis of Itti and Koch's [IKN98] traditional visual attention model. Image from [IK01a] reprinted with permission from NPG

putational implementation of this model was proposed by Koch and Ullman [KU85], who also coined the term "saliency map" that is identical to the concept of Treisman's "master map of location". The general idea is to compute several features in parallel that are fused to form the final saliency map. This traditional structure consists of several processing steps to calculate the saliency map and the different computational models differ in how they implement these steps. For example, Frintrop's visual object detection with computational attention system (VOCUS) uses integral images to calculate the center-surround differences [Fri06], Harel et al.'s graph-based visual saliency model [HKP07] implements Itti and Koch's model [IKN98], which is depicted in Fig. 2.3, in a consistent graph-based framework.

In this model, one or several image pyramids are computed to facilitate the subsequent computation features are computed on different scales. Then, image features are computed, which typically are based on local contrast operations such as, most importantly, "center-surround differences" that compare the average value of a center region with the average value in the surrounding region [Mar82]. The most common low-level feature channels are intensity, color, orientation, and motion. Each feature channel is subdivided into several feature types such as, for example, red, green, blue, and yellow feature maps for color. The features are commonly represented in so-called "feature maps", which are also known as "conspicuity maps". These feature maps are then normalized and fused to form a single "saliency map".

How the conspicuity maps are fused is a very important aspect of attention models. It is important that image regions that stand out in one feature map are not suppressed by the other feature maps. Furthermore, the feature calculation can be non-linear, leading to strong variations in the value range across and even within feature channels. Typical normalizations not just try to normalize the value range but also try to highlight local maxima and suppress the often considerable noise in the feature maps [IK01a, IKN98, Fri06]. The feature maps can be weighted, for example, bottom-up by their uniqueness or top-down to incorporate task knowledge when fused into the final saliency map.

Although the saliency map can serve as input to subsequent processing operations, e.g. as a relevance map for image regions, many applications require a trajectory of image regions similar to human saccades. Saccadic movement of the human eye is an essential part of the human visual system and critical to focus and resolve objects. By moving the eye, the small part of the scene that is fixated can be sensed with greater resolution, because it is projected on the central part of the retina, i.e. the fovea, which is responsible for highly resoluted, sharp, non-peripheral vision. To serially attend image regions, the saliency map's local maxima are determined and sequentially attended, typically in the order of descending saliency. A major contribution of Koch and Ullman [KU85] was to show that serially extracting the local maxima can be implemented with biologically-motivated winner-take-all (WTA) neural networks. To serially shift the focus of attention, the saliency of an attended region is suppressed so that the return of the focus of attention to previously attended regions is inhibited.

The computational model as described so far mostly reflects bottom-up attention, i.e. it does not explicitly handle task-specific top-down information (e.g. as is given by a sentence that describes a searched object such as "search for the red ball"). The

(a)   **(b)**   **(c)**

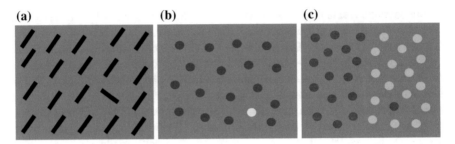

Fig. 2.4 Psychologically motivated test patterns that haven been and are still used to assess the capabilities of visual attention models [KF11]. The goal is to highlight the irregularities in the patterns. Images from and used with permission from Simone Frintrop. **a** Orientation. **b** Color. **c** Locality

most common approach to integrate top-down information is control the influence of the feature maps during the fusion and adapt the weights in such way that feature maps that are likely to highlight distractors are suppressed [Wol94, NI07]. The weights can either be static or dynamic to adapt the model to specific scenarios [XZKB10]. Additionally, it is possible to integrate further, more specialized feature maps that encode, for example, faces, persons, or even cars [JEDT09, CFK09].

Computational Models—Non-traditional Approaches

Since human eye movements are controlled by visual attention, which can easily be observed, gaze trajectories have long served as basis to study visual attention and aspects of human cognition in general. For example, in 1967, Yarbus showed that eye movements depend on the task that is given to a person [Yar67]. Consequently, the main goal of psychological models is to explain and predict eye movements that are recorded during eye tracking experiments. However, due to the lack of modern computerized eye tracking equipment, the abilities of visual attention models where for a long time assessed by testing whether or not they were able to replicate effects that have been observed on psychological test patterns, see Fig. 2.4. In the last five years, several eye tracking datasets have been made publicly available to evaluate visual attention models (e.g., [KNd08, BT09, CFK09, JEDT09]; Winkler and Subramanian provide an up-to-date overview of eye tracking datasets [WS13]). Among other aspects, such easily accessible datasets and the resulting quantitative comparability of test results has lead to a plethora of novel algorithms such as, for example, attention by information maximization [BT09], saliency using natural statistics [ZTM+08], graph-based visual saliency [HKP07], context-aware saliency [GZMT12, GZMT10], and Judd et al.'s machine learning model [JEDT09]. Interestingly, Borji et al. recently evaluated many proposed visual saliency algorithms on eye tracking data [BI13, BSI13b].

However, although being often evaluated on eye tracking data, most recently proposed models do not try to implement or explain any psychological or neurobiological models (e.g., [HHK12, HZ07]). However, a biological plausibility can sometimes be discovered later (e.g., [BZ09]). One such recent trend are spectral saliency models that were first proposed by Hou et al. [HZ07]. These models operate in the image's frequency spectrum and exploit the well-known effect that spectral whitening of signals will "accentuate lines, edges and other narrow events without modifying their position" [OL81]. Since these models are based on the fast Fourier transform (FFT), they combine state-of-the-art results in predicting where people look with the computational efficiency inherited from the FFT. Please note that spectral saliency models are discussed in detail in Sect. 3.1.1.

Another recent trend is to use machine learning techniques to learn to predict where humans look, which was first proposed by Judd et al. [JEDT09]. Most saliency models that rely on machine learning are either pixel- or patch-based. Pixel-based approaches have in common with the traditional structure of computational models that they calculate a collection of feature maps. Then, classification or regression methods such as, for example, support vector machines [JEDT09] or boosting [Bor12] can be trained to learn how to optimally fuse the individual feature maps into the final saliency map. Patch-based approaches compare image patches against each other to calculate the saliency of each patch. For example, it is possible to rank the image patches by their uniqueness and assign a high saliency to patches that contain features that are rarely seen across the image [LXG12]. However, all approaches that rely on machine learning have the disadvantage that they require enough training data, which can be problematic, because most datasets consist of a very limited number of eye tracked images.

Computational Models—Salient Object Detection

Recently, Liu et al. adapted the traditional definition of visual saliency by incorporating the high level concept of a salient object into the process of visual attention computation [LSZ+07]. A "salient object" is defined as being the object in an image that attracts most of the user's interest such as, for example, the man, the cross, the baseball players and the flowers that are shown in Fig. 2.5. Accordingly, Liu et al.

Fig. 2.5 Example images from Achanta et al.'s and Liu et al.'s salient object detection dataset [AS10, LSZ+07]

[LSZ+07] defined the task of "salient object detection" as the binary labeling problem of separating the salient object from the background. Here, it is important to note that the selection of a salient object happens consciously by the user whereas the gaze trajectories, which are recorded with eye trackers, are the result of mostly unconscious processes. Consequently, considering that salient objects naturally attract human gaze [ESP08], salient object detection and predicting where people look are very closely related yet different tasks with different evaluation measures and characteristics.

Since the ties of salient object detection to psychology and neurobiology are relatively loose, a wide variety of models has been proposed in recent years that are even less restricted by biological principles than traditional visual saliency algorithms. Initially, Liu et al. [LSZ+07] combined multi-scale contrast, center-surround histograms, and color spatial-distributions with conditional random fields. Liu et al.'s ideas—a combination of histograms, segmentation, and machine learning—can still be found in most salient object detection algorithms. Alexe et al. [ADF10] combine traditional bottom-up saliency, color contrast, edge density, and superpixels in a Bayesian framework. Closely related to Bayesian surprise [IB06], Klein et al. [KF11] use the Kullback-Leibler divergence of the center and surround image patch histograms to calculate the saliency map, whereas Lu and Lim [LL12] calculate and invert the whole image's color histogram to predict the salient object. Achanta et al. [AHES09, AS10] rely on the difference of pixels to the average color and intensity value of an image patch or even the whole image. Cheng et al. [CZM+11] use segmentation and define each segments saliency based on the color difference and spatial distance to all other segments.

2.1.2 Auditory Attention

Auditory attention is an important, complex system of bottom-up—i.e., sound-based salience—and top-down—i.e., task-dependent—aspects. Among other aspects, auditory attention assists in the computation of early auditory features and acoustic scene analysis,[1] the identification and recognition of salient acoustic objects, enhancement of signal processing for the attended features or objects, and the planning of actions in response to incoming auditory information [FEDS07]. Moreover, auditory attention can be directed to a rich set of acoustic features including, among others, spatial location, auditory pitch, frequency or intensity, tone duration, timbre, speech versus non-speech, and characteristics of individual voices [FEDS07]. The best example for these abilities is the "cocktail party effect" [Che08], which illustrates that we are able to attend and selectively listen to different speakers in a crowded room that is filled with a multitude of ongoing conversations. Consequently, auditory attention influences many levels of auditory processing; ranging from processing in the cochlea to the association cortex. Not unlike the "what" and "where" pathways in

---

[1] Auditory scene analysis describes the process of segregating and grouping sounds from a mixture of sources to determine and represent relevant auditory streams or objects [Bre90].

the human brain's visual system, there seem to be auditory "what" and "where" pathways, whose activation depends on whether an auditory task requires attending to an auditory feature or object or to a spatial location [ABGA04, DSCM07].

However, since auditory attention is an active research field in neurobiology, psychophysics, and psychology, it is only possible to provide a brief overview of selected aspects in the following. There exist however two detailed literature overviews: First, Hafter et al.'s review [HSL07] focuses on bottom-up aspects of auditory attention. Second, Fritz et al.'s survey [FEDS07] nicely presents aspects of top-down auditory and crossmodal attention. However, although there exists a large body of existing work, it is important to say that there are still many open research questions [FEDS07]. Some of these questions are directly related to the work presented in this book such as, for example: How much of the brain's acoustic novelty detection mechanisms can be explained by simple habituation mechanisms? What are the differences and similarities between visual and auditory attention? What is an appropriate computational model of auditory attention?

How Humans Perceive Sound

As shown in Fig. 2.6b, the cochlea is a coiled system of three ducts: the vestibular duct (scala vestibuli, upper gallery), the tympanic duct (scala tympani, lower gallery), and the cochlear duct (scala media, middle gallery). All of which are filled with lymphatic fluid. The cochlea contains a partition which is known as the "basilar membrane", see Fig. 2.7. The basilar membrane is essential for our sense of hearing and consists of, most importantly, the scala media, the organ of Corti, the tectorial membrane, and the basilar membrane.

Sound waves that reach the ear lead to oscillatory motions of the auditory ossicles. The oval window allows the transmission of this stimulus into the cochlea. In the cochlea, this stimulus sets the basilar membrane as well as the fluids in the scalae vestibuli and tympani in motion. The location of the maximal amplitude of the travelling wave that moves the basilar membrane depends on the frequency of the incoming sound signal. In other words, the basilar membrane performs a frequency analysis of the incoming sound wave. The motion along the basilar membrane stimulates nerve cells that are located in the organ of Corti, see Fig. 2.7. These nerve cells send electrical signals to the brain, which are finally perceived as sound.

Bottom-Up Auditory Attention

As mentioned before, attentional effects in the human auditory system can occur at various levels of auditory processing. Interestingly, the earliest, mostly bottom-up attentional mechanisms can be observed already in the cochlea [FEDS07, DEHR07, HPSJ56].

The ability to detect "novel", "odd", or "deviant" sounds amidst the environmental background noise is an important survival skill of humans and animals. Accordingly, the brain has evolved a sophisticated system to detect novel, odd, and deviant sounds. This system includes an automatic, pre-attentive component that analyzes stability and novelty of the acoustic streams within the acoustic scene, even for task-irrelevant acoustic streams [FEDS07, WTSH+03, WCS+05, Sus05].

(a)

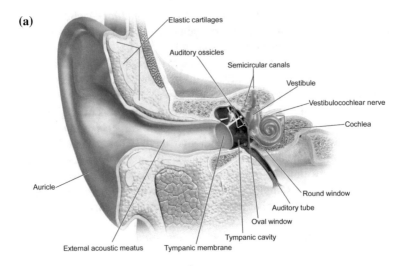

(b)

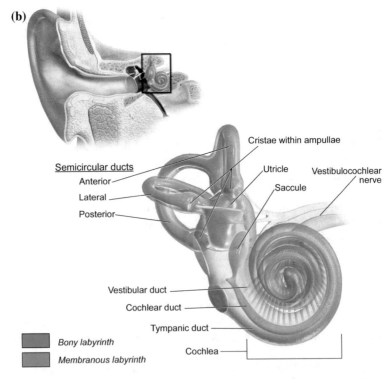

Fig. 2.6 Images illustrating the structure and transmission of sounds in the human ear. Images from [Bla14]. **a** Structure of the human ear [Wika]. **b** Structure of the human inner ear [Wikb]

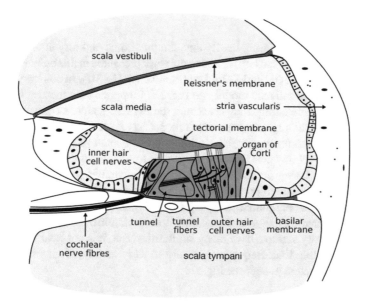

Fig. 2.7 Crosssection of the cochlea. Image from [Wikd].

The brain's acoustic novelty detection system consists of an interconnected set of mechanisms, which includes "adaptive" neurons and a specialization of so-called "novelty" detection neurons. Here, novelty detection neurons specifically encode deviations from the pattern of preceding stimuli. There exist two alternative views on this "change detection" within the auditory scene, depending on where the triggered novelty responses arise in the brain. According to the first view, novelty signals can occur very early in the human auditory system [PGMC05] and suggest the possibility of subcortical pathways for change detection [FEDS07]. However, most research focuses on projections of current neural sound representations that are matched against incoming sounds [FEDS07]. In this view, the change detection system continuously monitors the auditory environment, tracks changes, and updates its representation of the acoustic scene [SW01]. Here, the matching and the novelty response is a largely pre-attentive mechanism, which however can be influenced by top-down mechanisms. It has been shown that this kind of signal mismatch detection can be triggered by deviations in stimulus frequency, intensity, duration or spatial location, or by irregularities in spectrotemporal sequences (over periods of up to 20 s), or even in patterns of complex sounds such as speech and music [FEDS07]. Once such a novel or odd stimulus is detected and marked, it can be analyzed by the auditory system to decide whether it should receive further attention or even trigger a behavioral response. Unfortunately, the exact neural basis of this impressive fast, pre-attentive change detection system has not conclusively been found so far.

Computational Models

In contrast to visual attention, hardly any computational auditory attention models exist (cf. [Kal09]). Most closely related to the work presented in this book is the model by Kayser et al. [KPLL05] and Kalinli and Narayanan [KN07]. In both models, Itti et al.'s [IKN98] visual saliency model, see Fig. 2.3, is applied on a map representation of the acoustic signal's frequency spectrum, see Fig. 2.8, which is equivalent or very similar to the signal's spectrogram. This model has been successfully applied and extended for speech processing by Kalinli and Narayanan [KN09, Kal09, KN07], where Kalinli and Narayanan focus on integrating top-down influences in the auditory attention model. Using the auditory spectrum of incoming sound as the basis for bottom-up auditory attention mimics "the process from basilar membrane to the cochlear nucleus in the human auditory system" [Kal09]. Transferring Itti et al.'s visual model to auditory signals is a radical implementation of the idea that the human visual and auditory systems have many similarities. But, it is not clear whether there exists an accessible time-frequency memory in early audition as is implied by the model's time-frequency map, see Fig. 2.8.

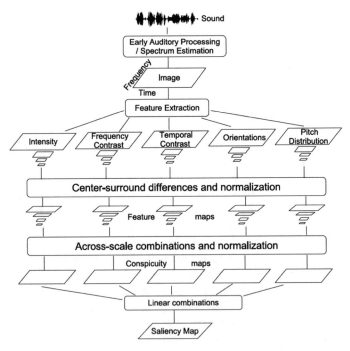

Fig. 2.8 Itti and Koch's visual saliency model [IKN98] transferred to auditory saliency detection as has been proposed by Kayser et al. [KPLL05] and similarly by Kalinli and Narayanan [KN07]. Image from [KN07] with permission from ISCA

2.1.3 Multimodal Attention

Crossmodal Integration

There exist substantial similarities between the visual and auditory attention systems: Most importantly, both consist of bottom-up and top-down components and there appear to be specialized "what" and "where" processes. Since a few years, there is an increasing amount of experimental results that show that all sensory processing in the human brain is in fact multisensory [GS06]. For example, it has been shown that lipreading [CBB+97] or the observation of piano playing without hearing the sound [HEA+05] can activate areas in the auditory cortex.

Several studies have shown that the presence of a visual stimulus or attending a visual task can draw away attention from an auditory stimulus, which is indicated by a decreased activity in the auditory cortex (e.g., [LBW+02, WBB+96]). Similarly, auditory attention can negatively influence visual attention. In fact, it was shown that there exists a reciprocal inverse relationship between auditory and visual activation, which means that increases in visual activation correlate with a decrease in auditory activation and vice versa. A very interesting study was performed by Weissman et al. [WWW04]. Weissman et al. created a conflict between auditory and visual target stimuli, and crossmodal distractors. They observed that when the "distracting stimulus in the task-irrelevant sensory channel is increased, there was a compensatory increase in selective attention to the target in the relevant channel and a corresponding increase in activation in the relevant sensory cortex" [FEDS07]. This suggests that it is likely that there exists a top-down mechanism that regulates the relative strengths of the sensory channels.

How auditory and visual sensory information interact for the control of overt attention, i.e. directing the sensory organs toward interesting stimuli, has recently been investigated by Onat et al. [OLK07]. Onat et al. performed eye tracking studies in which the participants were listening to sounds coming from different directions. It was shown that eye fixation probabilities increase toward the location where the sound originates, which means—unsurprisingly—that the selection of fixation points depends on auditory and visually salient stimuli. Furthermore, Onat et al. used the data to test several biologically plausible crossmodal integration mechanisms and found "that a linear combination of both unimodal saliencies provides a good model for this integration process" [OLK07]. Interestingly, such a linear combination is not just optimal in an information theoretic sense (see [OLK07]), but it also allows to adjust the relative strength of the sensory channels. However, this model assumes the existence of a 2-dimensional auditory saliency map that encodes where salient stimuli occur in the scene and how salient these stimuli are, which can be directly fused with the visual saliency map to form a joint audio-visual saliency map.

High-Level Influences

Not just crossmodal effects can influence what is interesting in one modality. Instead, there exist many top-down signals that can direct attention toward specific targets (e.g., [Ban04, Hob05, CC03, TTDC06, STET01, WHK+04, NI07]).

Verbal descriptions of object properties can directly influence what is perceived as being perceptually salient (e.g., [STET01, WHK+04, NI07]). For example, it has been shown that knowledge about an object's visual appearance can influence the perceptual saliency to highlight an object that we are actively searching, i.e. in a so-called visual search task. But, only specific information that refers to primitive preattentive features allows such attentional guidance [WHK+04]. Accordingly, if we have a good visual impression or memory of the target that we are looking for (e.g., we have just seen it a few moments ago) or if we at least know the target's color, then we can find the target faster. In theses cases, the visual saliency will be guided in such way that it stronger highlights image regions that exhibit the target's preattentive features. In contrast, for example, categorical information about the search target (e.g., search for an animal) typically does not provide such top-down guidance (see [WHK+04]).

Interestingly, certain features that would typically be associated with high-level vision tasks can attract our low-level attention independent of task. Most importantly, it has be shown that faces and face-like patterns attract the gaze of infants as young as 6 weeks, i.e. before they can consciously perceive the category of faces [SS06]. The fact that the gaze and, consequently, interest of infants is attracted by face-like patterns seems to be an important aspect of early infant development, especially for social signals and processes (see, e.g., [KJS+02]). Interestingly, infants show the ability to follow the observed gaze direction of caregivers at an age of 6 months [Hob05]. If people talk about objects that are part of the environment, where and at what people are looking at is related to the object that is being talked about. Consequently, the ability to follow the caregiver's gaze makes it possible for an infant to associate what it sees with the words it hears, an important ability to learn a language. Similar to gaze but more direct and less subtle, infants also soon develop the ability to interpret pointing gestures (see [LT09]). Accordingly, pointing gestures and gaze are both non-verbal signals that direct the attention toward a spatial region of interest (see, e.g., [Ban04, LB05, LT09]). This is an essential aspect in natural interaction, because it makes it possible to direct and coordinate the attention of interacting persons and, thus, helps to establish a joint focus of attention. In other words, such non-verbal signals are used to influence where an interaction partner is looking in order to direct his gaze toward a specific object that is or will become the subject of the conversation. Consequently, the generation and interpretation of such signals is fundamental for "learning, language, and sophisticated social competencies" [MN07a].

2.2 Applications of Attention Models

Knowing in advance what people might find interesting and attend to is an important information that can be integrated into many applications. Images and videos can be compressed better, street signs can be designed to immediately grab the attention, and advertisement can put stronger emphasis on the intended message. Furthermore, having an estimate of what is probably a relevant signal in a data stream allows us to focus computational algorithms. This way, machine learning can learn better models from less data, class-independent object detection as well as object recognition can be improved, and robots are able to process incoming sensory information in real-time despite limited computational power.

2.2.1 Image Processing and Computer Vision

Image and video compression algorithms can improve the perceptual quality of compressed images and videos by allocating more bits to code image regions that exhibit a high perceptual saliency [GZ10, OBH+01]. This way, image regions that are likely to attract the viewers' interest are less compressed and thus show fewer disturbing alterations such as compression artifacts. Ouerhani et al. [OBH+01] implement such an adaptive coding scheme that favors the allocation of a higher number of bits to those image regions that are more conspicuous to the human visual system. The compressed image files are fully compatible with the JPEG standard. An alternative approach was recently proposed by Hadizadeh and Bajic [HB13]. Their method uses saliency to automatically reduce potentially attention-grabbing coding artifacts in regions of interest.

Visual attention and object recognition are tightly linked processes in human perception. Accordingly, although most models of visual attention and object recognition are separated, there is an increasing interest in integrating both processes to increase the performance of computer vision systems. Initial approaches tried to use attention as a front-end to detect salient objects or keypoint locations. Miau et al. [MPI01] use an attentional front-end with the biologically motivated object recognition system HMAX [RP99]. Walther and Koch [WK06] combine an attention system with a SIFT-based object recognition [Low04] and demonstrate that they are able to improve object recognition performance. Going a step further, Walter and Koch [WK06] suggest a unifying attention and object recognition framework. In this framework, the HMAX object recognition is modulated to suppress or enhance image locations and features depending on the spatial attention.

Related to such attentional front-ends for object recognition, principles of visual attention have recently been integrated into approaches for general, class independent object detection [ADF10]. This way, sampling windows can be distributed according to the "objectness" distribution and used as location priors for class-specific object detectors. This can greatly reduce the necessary number of windows evaluated by class-specific object detectors as has been shown in the PASCAL Visual Object Classes (VOC) challenge 2007 [EVGW+]. Interestingly, going in the other direction, high-level object detectors are being integrated into saliency models to model, for example, that the human visual attention is attracted by faces and face-like patterns. For this purpose, some models integrate detectors for faces, the horizon, persons and even cars [CHEK07, JEDT09]. This shows that attention and object recognition might grow together in the future.

Saliency has also been employed as a spatial prior to learn object attributes, categories, or classes from weakly labeled images. For example, Fei-Fei Li et al.'s

[FFFP03] approach to "one-shot learning" uses Kadir and Brady's saliency detector [KB01] to sample features at highly salient locations. The most salient regions are clustered over location and scale to give a reasonable number of distinctive features per image.

2.2.2 Audio Processing

In contrast to applications in computer vision, only few applications of acoustic or auditory saliency models have been explored so far. Coensel and Botteldooren [CB10] propose to use an auditory attention model in soundscape design to assess how specific sounds can mask unwanted environmental sounds. Lin et al. [LZG+12] as well as, in principle, Kalinli and Narayanan [KN07, KN09] use Itti's classic visual saliency model [IKN98] to highlight visually salient patterns in the spectrogram. Lin et al. [LZG+12] fuse the spectrogram's saliency map with the original spectrogram and use the resulting saliency-maximized audio spectrogram to enable faster than real-time detection of audio events by human audio analysts. Kalinli and Narayanan [KN07] use the spectrogram's saliency map to detect prominent syllable and word locations in speech, achieving close to human performance. The task of syllable detection was chosen by the authors to investigate low-level auditory saliency models, because during speech perception, a particular phoneme or syllable can be perceived to be more salient than the others due to the coarticulation between phonemes, and other factors such as the accent, and physical and emotional state of the talker [KN07].

2.2.3 Robotics

In addition to reducing computational requirements by focusing on the most salient stimuli, robots can benefit from attentional mechanisms at several conceptual levels [Fri11]. On a low level, attention can be used for salient landmark detection and subsequent scene recognition and localization. On a mid level, attention can serve as a pre-processing step for object recognition. On a high level, attention can be implemented in a human-like fashion to guide actions and mimic human behavior, for example, during object manipulation or human-robot interaction.

Salient landmarks are excellent candidates for localization, because they are visually outstanding and distinctive, often having unique features. This makes them easy to (re-)detect and allows for a very sparse set of localization landmarks that can easily be detected, accessed in memory, and matched in real-time. The ARK project [NJW+98] is one of the earliest projects that investigated the use of salient landmarks for localization. The localization was based on manually generated maps of static obstacles and natural visual landmarks. Siagian and Itti [SI09] presented an integrated system for coarse global localization based on the "gist" of the scene and fine localization within a scene using salient landmarks. Frintrop and Jensfelt [FJ08] combined attention and salient landmark detection with simultaneous localization

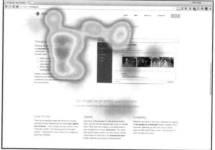

Fig. 2.9 Eye tracking experiments are used to optimize the layout of websites. The opacity map (*left*) illustrates what people see and the heat map (*right*) represents the distribution of interest on the website. Images generated with Eyezag's online eye tracking platform [Eye]

and mapping (SLAM). The attention system VOCUS [Fri06] detects salient regions. These regions are tracked and matched to all entries in a database of previously seen landmarks to estimate a 3D position.

The main difference between robotic applications and, for example, image processing is that a robot can move its body parts to interact with its environment and influence what it perceives. This way, robots can control their geometric parameters, e.g. where it looks, and manipulate the environment to improve the perception quality of specific stimuli [AWB88]. This can be implemented with an attentive two-step object detection and recognition mechanism: First, regions of interest are detected in a peripheral vision system based on visual saliency and a coarse view of the scene. Second, the robot then investigates each region of interest by focusing its sensors on the target object, which provides high-resolution images for object recognition (e.g., [MFL+08, GAK+07]). It is noteworthy to say that using this strategy Meger et al.'s robot "Curious George" [MFL+08] won the 2007 and 2008 Semantic Robotic Vision Challenge [Uni] (Fig. 2.9).

A common assumption in the field of socially interactive robots is that "humans prefer to interact with machines in the same way that they interact with other people" [FND03]. This is based on the observation that humans tend to treat robots like people and, as a consequence, tend to expect human-like behavior from robots [FND03, NM00]. According to this assumption, a computational attention system that mimics how humans direct their attention can facilitate human-robot interaction. For example, this idea has been implemented in the social robot Kismet, whose gaze is controlled by a visual attention system [BS99].

2.2.4 Computer Graphics

Naturally, knowing what attracts the viewer's attention is important when automatically generating or manipulating images. For example, it is possible to automatically crop an image to only present the most relevant content to a user and/or act as a thumbnail [SLBJ03, SAD+06, CXF+03]. Similarly, content-aware media retargeting automatically changes the aspect ratio of images and videos to optimize the presentation of visual content across platforms and screen sizes [SLNG07, AS07, RSA08, GZMT10]. For this purpose, saliency models are used to automatically determine image regions that are likely to contain relevant information. Depending on their estimated importance, image regions are then deleted or morphed so that the resized image best portrays the most relevant information (see, e.g., [SLNG07]).

2.2.5 Design, Marketing, and Advertisement

There exist several companies such as, for example, SMIvision [SMI] and Eyezag [Eye] that offer eye tracking experiments as a service. This enables companies to analyze how people view their webpage, advertisement, or image and video footage, see Figs. 2.9 and 2.10. Other companies such as Google have their own in-house laboratories and solutions to perform eye tracking experiments and research [Goo].

With increasingly powerful computational attention models that predict human fixations, it becomes possible to reduce the need for expensive and intrusive eye tracking experiments. In 2013, 3M has started to offer its visual attention service [AMR3M] that uses a computational attention model as a cheaper and faster alternative to eye tracking experiments. Potential usage scenarios as proposed by 3M are in-store merchandising, packaging, advertising, web and banner advertisement, and video analysis [AMR3M].

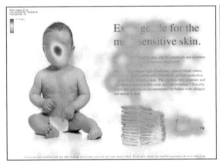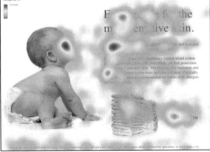

Fig. 2.10 Eye tracking experiments are used to optimize the layout of advertisement. Images from and used with permission from James Breeze

References

[AHES09]   Achanta, R., Hemami, S., Estrada, F., Süsstrunk, S.: Frequency-tuned salient region detection. In: Proceedings of the International Conference on Computer Vision Pattern Recognition (2009)

[AS10]   Achanta, R., Süsstrunk, S.: Saliency detection using maximum symmetric surround. In: Proceedings of the International Conference on Image Processing (2010)

[ADF10]   Alexe, B., Deselaers, T., Ferrari, V.: What is an object? In: Proceedings of the International Conference on Computer Vision Pattern Recognition, pp. 73–80 (2010)

[AWB88]   Aloimonos, Y., Weiss, I., Bandopadhay, A.: Active vision. Int. J. Comput. Vis. **1**(4), 333–356 (1988)

[ABGA04]   Arnott, S.R., Binns, M.A., Grady, C.L., Alain, C.: Assessing the auditory dual-pathway model in humans. Neuroimage **22**, 401–408 (2004)

[AS07]   Avidan, S., Shamir, A.: Seam carving for content-aware image resizing. ACM Trans. Graph. **26**(3) (2007)

[AMR3M]   3SMM: 3M visual attention service. http://solutions.3m.com/wps/portal/3M/en_US/VAS-NA?MDR=true

[Ban04]   Bangerter, A.: Using pointing and describing to achieve joint focus of attention in dialogue. Psychol. Sci. **15**(6), 415–419 (2004)

[BZ09]   Bian, P., Zhang, L.: Biological plausibility of spectral domain approach for spatiotemporal visual saliency. In: Proceedings of the Annual Conference on Neural Information Processing Systems (2009)

[Bla14]   Blausen.com staff: Blausen gallery 2014. Wikiversity J. Med. (2014)

[Bor12]   Borji, A.: Boosting bottom-up and top-down visual features for saliency estimation. In: Proceedings of the International Conference on Computer Vision Pattern Recognition (2012)

[BI13]   Borji, A., Itti, L.: State-of-the-art in visual attention modeling. IEEE Trans. Pattern Anal. Mach. Intell. **35**(1), 185–207 (2013)

[BSI13b]   Borji, A., Sihite, D.N., Itti, L.: Quantitative analysis of human-model agreement in visual saliency modeling: a comparative study. IEEE Trans. Image Process. **22**(1), 55–69 (2013)

[BH05]   Breazeal, C., Scassellati, B.: A context-dependent attention system for a social robot. In: Proceedings of the International Joint Conference on Artificial Intelligence (1999)

[BS99]   Bregman, A.S.: Auditory Scene Analysis: The Perceptual Organization of Sounds. MIT Press (1990)

[Bre90]   Bruce, N., Tsotsos, J.: Saliency, attention, and visual search: an information theoretic approach. J. Vis. **9**(3), 1–24 (2009)

[BT09]   Bundesen, C., Habekost, T.: Handbook of Cognition. Sage Publications, Chap. Attention (2005)

[CBB+97]   Calvert, G.A., Bullmore, E., Brammer, M., Campbell, R., Williams, S.C., McGuire, P.K., Woodruff, P.W., Iversen, S.D., David, A.S.: Activation of auditory cortex during silent lipreading. Science **276**, 593–596 (1997)

[CC03]   Cashon, C., Cohen, L.: The Construction, Deconstruction, and Reconstruction of Infant Face Perception. Chapter The development of face processing in infancy and early childhood, Current perspectives, pp. 55–68. NOVA Science Publishers (2003)

[CHEK07]   Cerf, M., Harel, J., Einhäuser, W., Koch, C.: Predicting human gaze using low-level saliency combined with face detection. In: Proceedings of the Annual Confernce on Neural Information Processing Systems (2007)

[CFK09]   Cerf, M., Frady, E.P., Koch, C.: Faces and text attract gaze independent of the task: experimental data and computer model. J. Vis. **9** (2009)

[CXF+03]   Chen, L.-Q., Xie, X., Fan, X., Ma, W.-Y., Zhang, H.-J., Zhou, H.-Q.: A visual attention model for adapting images on small displays. Multim. Syst. **9**(4), 353–364 (2003)

[CZM+11]   Cheng, M.-M., Zhang, G.-X., Mitra, N.J., Huang, X., Hu, S.-M.: Global contrast
           based salient region detection. In: Proceedings of the International Conference on
           Computer Vision Pattern Recognition (2011)
[Che08]   Cherry, E.C.: Some experiments on the recognition of speech, with one and with two
           ears. J. Acoust. Soc. Am. **25**, 975–979 (2008)
[CB10]   Coensel, B.D., Botteldooren, D.: A model of saliency-based auditory attention to
           environmental sound. In: Proceedings of the International Congress on Acoustics
           (2010)
[DEHR07]   Delano, P.H., Elgueda, D., Hamame, C.M., Robles, L.: Selective attention to visual
           stimuli reduces cochlear sensitivity in chinchillas. J. Neurosci. **27**, 4146–4153 (2007)
[DSCM07]   De Santis, L., Clarke, S., Murray, M.M.: Automatic and intrinsic auditory what and
           where processing in humans revealed by electrical neuroimaging. Cereb Cortex **17**,
           9–17 (2007)
[ESP08]   Einhäuser, W., Spain, M., Perona, P.: Objects predict fixations better than early
           saliency. J. Vis. **8**(14) (2008)
[EVGW+]   Everingham, M., Van Gool, L., Williams, C.K.I., Winn, J., Zisserman, A.: The PAS-
           CAL Visual Object Classes Challenge 2007 (VOC2007) Results. http://www.pascal-
           network.org/challenges/VOC/voc2007/workshop/index.html
[Eye]   Eyezag: Eyezag—eye tracking in your hands. http://www.eyezag.com/
[FFFP03]   Fei-Fei, L., Fergus, R., Perona, P.: A bayesian approach to unsupervised one-shot
           learning of object categories. In: Proceedings of the International Conference on
           Computer Vision (2003)
[FND03]   Fong, T., Nourbakhsh, I., Dautenhahn, K.: A survey of socially interactive robots.
           Robot. Auton. Syst. **42**(3–4), 143–166 (2003)
[Fri06]   Frintrop, S.: VOCUS: A Visual Attention System for Object Detection and Goal-
           Directed Search, ser. Springer, Lecture Notes in Computer Science (2006)
[FJ08]   Frintrop, S., Jensfelt, P.: Attentional landmarks and active gaze control for visual
           slam. IEEE Trans. Robot. **24**(5), 1054–1065 (2008)
[FRC10]   Frintrop, S., Rome, E., Christensen, H.I.: Computational visual attention systems and
           their cognitive foundation: a survey. ACM Trans. Applied Percept. **7**(1), 6:1–6:39
           (2010)
[FEDS07]   Fritz, J.B., Elhilali, M., David, S.V., Shamma, S.A.: Auditory attention-focusing the
           searchlight on sound. Curr. Opin. Neurobiol. **17**(4), 437–455 (2007)
[GS06]   Ghazanfar, A.A., Schroeder, C.E.: Is neocortex essentially multisensory? Trends
           Cogn. Sci. **10**, 278–285 (2006)
[GZMT10]   Goferman, S., Zelnik-Manor, L., Tal, A.: Context-aware saliency detection. In: Pro-
           ceedings of the International Conference on Computer Vision Pattern Recognition
           (2010)
[Goo]   Google: Eye-tracking studies: more than meets the eye. http://googleblog.blogspot.
           de/2009/02/eye-tracking-studies-more-than-meets.html
[GAK+07]   Gould, S., Arfvidsson, J., Kaehler, A., Sapp, B., Messner, M., Bradski, G., Baum-
           starck, P., Chung, S., Ng, A.Y.: Peripheral-foveal vision for real-time object recogni-
           tion and tracking in video. In: Proceedings of the International Joint Conference on
           Artificial Intelligence (2007)
[GZ10]   Guo, C., Zhang, L.: A novel multiresolution spatiotemporal saliency detection model
           and its applications in image and video compression. IEEE Trans. Image Process. **19**,
           185–198 (2010)
[HB13]   Hadizadeh, H., Bajic, I.: Saliency-aware video compression. IEEE Trans. Image
           Process (99) (2013)
[HSL07]   Hafter, E.R., Sarampalis, A., Loui, P.: Auditory Perception of Sound Sources. Springer
           (2007) (ch. Auditory attention and filters (review))
[HEA+05]   Haslinger, B., Erhard, P., Altenmuller, E., Schroeder, U., Boecker, H., Ceballos-
           Baumann, A.O.: Transmodal sensorimotor networks during action observation in
           professional pianists. J. Cogn. Neurosci. **17**, 282–293 (2005)

[HKP07]   Harel, J., Koch, C., Perona, P.: Graph-based visual saliency. In: Proceedings of the Annual Conference on Neural Information Processing Systems (2007)

[HPSJ56]   Hernandez-Peon, R., Scherrer, H., Jouvet, M.: Modification of electric activity in cochlear nucleus during attention in unanesthetized cats. Science **123**, 331–332 (1956)

[Hob05]   Hobson, R.: Joint attention: Communication and other minds. Oxford University Press (2005) (Chap.. What puts the jointness in joint attention?), pp. 185–204

[HHK12]   Hou, X., Harel, J., Koch, C.: Image signature: highlighting sparse salient regions. IEEE Trans. Pattern Anal. Mach. Intell. **34**(1), 194–201 (2012)

[HZ07]   Hou, X., Zhang, L.: Saliency detection: a spectral residual approach. In: Proceedings of the International Conference on Computer Vision Pattern Recognition (2007)

[IB06]   Itti, L., Baldi, P.: Bayesian surprise attracts human attention. In: Proceedings of the Annual Confernce on Neural Information Processing Systems (2006)

[IKN98]   Itti, L., Koch, C., Niebur, E.: A model of saliency-based visual attention for rapid scene analysis. IEEE Trans. Pattern Anal. Mach. Intell. **20**(11), 1254–1259 (1998)

[IK01a]   Itti, L., Koch, C., Niebur, E.: Computational modelling of visual attention. Nat. Rev. Neurosci. **2**(3), 194–203 (2001)

[JSF12]   Jaspers, H., Schauerte, B., Fink, G.A.: Sift-based camera localization using reference objects for application in multi-camera environments and robotics. In: Proceedings of the 1st International Conference on Pattern Recognition Applications and Methods (ICPRAM), Vilamoura, Algarve, Portugal (2012)

[JEDT09]   Judd, T., Ehinger, K., Durand, F., Torralba, A.: Learning to predict where humans look. In: Proceedings of the International Conference on Computer Vision (2009)

[KB01]   Kadir, T., Brady, M.: Saliency, scale and image description. Int. J. Comput. Vis. **45**(2), 83–105 (2001)

[KTG92]   Kahneman, D., Treisman, A., Gibbs, B.J.: The reviewing of object files: object-specific integration of information. Cogn. Psychol. **24**(2), 175–219 (1992)

[Kal09]   Kalinli, O.: Biologically inspired auditory attention models with applications in speech and audio processing. Ph.D. dissertation, University of Southern California, Los Angeles, CA, USA (2009)

[KPLL05]   Kayser, C., Petkov, C.I., Lippert, M., Logothetis, N.K.: Mechanisms for allocating auditory attention: an auditory saliency map. Curr. Biol. **15**(21), 1943–1947 (2005)

[KN09]   Kalinli, O.: Prominence detection using auditory attention cues and task-dependent high level information. IEEE Trans. Audio Speech Lang Proc. **17**(5), 1009–1024 (2009)

[KN07]   Kalinli, O., Narayanan, S.: A saliency-based auditory attention model with applications to unsupervised prominent syllable detection in speech. In: Proceedings of the Annual Confernce on International Speech Communication Association (2007)

[KF11]   Klein, D.A., Frintrop, S.: Center-surround divergence of feature statistics for salient object detection. In: Proceedings of the International Conference on Computer Vision (2011)

[KJS+02]   Klin, A., Jones, W., Schultz, R., Volkmar, F., Cohen, D.: Visual fixation patterns during viewing of naturalistic social situations as predictors of social competence in individuals with autism. Arch. Gen. Psychiatry **59**(9), 809–816 (2002)

[KU85]   Koch, C., Ullman, S.: Shifts in selective visual attention: towards the underlying neural circuitry. Hum. Neurobiol. **4**, 219–227 (1985)

[KSS13]   Koester, D., Schauerte, B., Stiefelhagen, R.: Accessible section detection for visual guidance. In: IEEE/NSF Workshop on Multimodal and Alternative Perception for Visually Impaired People (2013)

[KNd08]   Kootstra, G., Nederveen, A., de Boer, B.: Paying attention to symmetry. In: Proceedings of the British Conference on Computer Vision (2008)

[KSSK12]   Kühn, B., Schauerte, B., Stiefelhagen, R., Kroschel, K.: A modular audio-visual scene analysis and attention system for humanoid robots. In: Proceedings of the 43rd International Symposium on Robotics (ISR), Taipei, Taiwan (2012)

[KSKS12]   Kühn, B., Schauerte, B., Kroschel, K., Stiefelhagen, R.: Multimodal saliency-based
            attention: A lazy robot's approach. In: Proceedings of the 25th International Con-
            ference on Intelligent Robots and Systems (IROS). IEEE/RSJ, Vilamoura, Algarve,
            Portugal (2012)

[SS13b]   Kühn, B., Schauerte, B., Kroschel, K., Stiefelhagen, R.: Wow! Bayesian surprise for
            salient acoustic event detection. In: Proceedings of the 38th International Conference
            on Acoustics, Speech, and Signal Processing (ICASSP). IEEE, Vancouver, Canada
            (2013)

[LBW+02]   Laurienti, P., Burdette, J.H., Wallace, M.T., Yen, Y.F., Field, A.S., Stein, B.E.: Deac-
            tivation of sensory-specific cortex by cross-modal stimuli. J. Cogn. Neurosci. **14**,
            420–429 (2002)

[LXG12]   Li, J., Xu, D., Gao, W.: Removing label ambiguity in learning-based visual saliency
            estimation. IEEE Trans. Image Process. **21**(4), 1513–1525 (2012)

[LT09]   Liebal, K., Tomasello, M.: Infants appreciate the social intention behind a pointing
            gesture: commentary on "children's understanding of communicative intentions in
            the middle of the second year of life" by T. Aureli, P. Perucchini and M. Genco.
            Cogn. Dev **24**(1), 13–15 (2009)

[LZG+12]   Lin, K.-H., Zhuang, X., Goudeseune, C., King, S., Hasegawa-Johnson, M., Huang,
            T.S.: Improving faster-than-real-time human acoustic event detection by saliency-
            maximized audio visualization. In: Proceedings of the International Confernce on
            Acoustics, Speech, and Signal Processing (2012)

[Fri11]   Lin, K.-H., Zhuang, X., Goudeseune, C., King, S., Hasegawa-Johnson, M., Huang,
            T.S.: Towards attentive robots. Paladyn **2**(2), 64–70 (2011)

[LSZ+07]   Liu, T., Sun, J., Zheng, N.-N., Tang, X., Shum, H.-Y.: Learning to detect a salient
            object. In: Proceedings of the International Conference on Computer Vision Pattern
            Recognition (2007)

[LL12]   Lu, S., Lim, J.-H.: Saliency modeling from image histograms. In: Proceedings of the
            European Confernce on Computer Vision (2012)

[LB05]   Louwerse, M., Bangerter, A.: Focusing attention with deictic gestures and linguistic
            expressions. In: Proceedings of the Annual Confernce on Cognitive Sciience Society
            (2005)

[Low04]   Lowe, D.G.: Distinctive image features from scale-invariant keypoints. Int. J. Comput.
            Vis. **60**, 91–110 (2004)

[Mar82]   Marr, D.: VISION—A Computational Investigation into the Human Representation
            and Processing of Visual Information. W.H Freeman and Company (1982)

[NI07]   Marr, D.: Search goal tunes visual features optimally. Neuron **53**(4), 605–617 (2007)

[MCS+14]   Martinez, M., Constantinescu, A., Schauerte, B., Koester, D., Stiefelhagen, R.: Cog-
            nitive evaluation of haptic and audio feedback in short range navigation tasks. In:
            Proceedings of the 14th Int. Conf. Computers Helping People with Special Needs
            (ICCHP). Springer, Paris, France (2014)

[MSS13]   Martinez, M., Schauerte, B., Stiefelhagen, R.: BAM! Depth-based body analysis
            in critical care. In: Proceedings of the 15th International Conference on Computer
            Analysis of Images and Patterns (CAIP). Springer, York, UK (2013)

[SS13a]   Martinez, M., Schauerte, B., Stiefelhagen, R.: How the distribution of salient objects
            in images influences salient object detection. In: Proceedings of the 20th International
            Conference on Image Processing (ICIP). IEEE, Melbourne, Australia (2013)

[MFL+08]   Meger, D., Forssén, P.-E., Lai, K., Helmar, S., McCann, S., Southey, T., Baumann, M.,
            Little, J.J., Lowe, D.J.: Curious george: an attentive semantic robot. Robot. Auton.
            Syst. **56**(6), 503–511 (2008)

[MPI01]   Miau, F., Papageorgiou, C., Itti, L.: Neuromorphic algorithms for computer vision
            and attention. In: Bosacchi, B., Fogel, D.B., Bezdek, J.C. (eds.) Proceedings of the
            SPIE 46 Annual International Symposium on Optical Science and Technology, vol.
            4479, pp. 12–23 (2001)

[MN07a]   Mundy, P., Newell, L.: Attention, joint attention, and social cognition. Curr. Dir. Psychol. Sci. **16**(5), 269–274 (2007)

[NM00]   Nass, C., Moon, Y.: Machines and mindlessness: social responses to computers. J. Soc. Issues **56**(1), 81–103 (2000)

[NJW+98]   Nickerson, S.B., Jasiobedzki, P., Wilkes, D., Jenkin, M., Milios, E., Tsotsos, J.K., Jepson, A., Bains, O.N.: The ark project: autonomous mobile robots for known industrial environments. Robot. Auton. Syst. **25**, 83–104 (1998)

[OLK07]   Onat, S., Libertus, K., König, P.: Integrating audiovisual information for the control of overt attention. J. Vis. **7**(10) (2007)

[OL81]   Oppenheim, A., Lim, J.: The importance of phase in signals. Proc. IEEE **69**(5), 529–541 (1981)

[OBH+01]   Ouerhani, N., Bracamonte, J., Hugli, H., Ansorge, M., Pellandini, F.: Adaptive color image compression based on visual attention. In: Proceedings of the International Conference on Image Analysis and Processing, pp. 416–421 (2001)

[PGMC05]   Perez-Gonzalez, D., Malmierca, M.S., Covey, E.: Novelty detector neurons in the mammalian auditory midbrain. Eur. J. Neurosci. **22**, 2879–2885 (2005)

[RP99]   Riesenhuber, M., Poggio, T.: Hierarchical models of object recognition in cortex. Nat. Neurosci. **2**, 1019–1025 (1999)

[RSA08]   Rubinstein, M., Shamir, A., Avidan, S.: Improved seam carving for video retargeting. In: Proceedings of the Annual Confernce on Special Interest Group on Graphics and Interactive Techniques (2008)

[RSAHS14] Rybok, L., Schauerte, B., Al-Halah, Z., Stiefelhagen, R.: Important stuff, everywhere! Activity recognition with salient proto-objects as context. In: Proceedings of the 14th IEEE Winter Conference on Applications of Computer Vision (WACV), Steamboat Springs, CO, USA (2014)

[SAD+06]   Santella, A., Agrawala, M., DeCarlo, D., Salesin, D., Cohen, M.: Gaze-based interaction for semi-automatic photo cropping. In: Proceedings of the International Conference on Human Factors Computing Systems (CHI) (2006)

[SF10a]   Schauerte, B., Fink, G.A.: Focusing computational visual attention in multi-modal human-robot interaction. In: Proceedings of the 12th International Conference on Multimodal Interfaces and 7th Workshop on Machine Learning for Multimodal Interaction (ICMI-MLMI). ACM, Beijing, China (2010)

[SF10b]   Schauerte, B., Fink, G.A.: Web-based learning of naturalized color models for human-machine interaction. In: Proceedings of the 12th International Conference on Digital Image Computing: Techniques and Applications (DICTA). IEEE, Sydney, Australia (2010)

[Sch14]   Schauerte, B.: Multimodal computational attention for scene understanding. Ph.D. dissertation, Karlsruhe Institute of Technology (2014)

[SKMS14]   Schauerte, B., Koester, D., Martinez, M., Stiefelhagen, R.: Way to Go! Detecting open areas ahead of a walking person. In: ECCV Workshop on Assistive Computer Vision and Robotics (ACVR). Springer (2014)

[SWS15b]   Schauerte, B., Koester, D., Martinez, M., Stiefelhagen, R.: A web-based platform for interactive image sonificationn. In: Accessible Interaction for Visually Impaired People (AI4VIP) (2015)

[SS14]   Schauerte, B., Koester, D., Martinez, M., Stiefelhagen, R.: Look at this! Learning to guide visual saliency in human-robot interaction. In: Proceedings of the International Conference on Intelligent Robots and Systems (IROS). IEEE/RSJ (2014)

[SS15]   Schauerte, B., Koester, D., Martinez, M., Stiefelhagen, R.: On the distribution of salient objects in web images and its influence on salient object detection. PLoS ONE **10**, 07 (2015)

[SKKS11]   Schauerte, B., Kühn, B., Kroschel, K., Stiefelhagen, R.: Multimodal saliency-based attention for object-based scene analysis. In: Proceedings of the 24th International Conference on Intelligent Robots and Systems (IROS). IEEE/RSJ, San Francisco, CA, USA (2011)

[SMCS12]   Schauerte, B., Martinez, M., Constantinescu, A., Stiefelhagen, R.: An assistive vision
           system for the blind that helps find lost things. In: Proceedings of the 13th International
           Conference on Computers Helping People with Special Needs (ICCHP). Springer,
           Linz, Austria (2012)
[SPF09]   Schauerte, B., Plötz, T., Fink, G.A.: 'A multi-modal attention system for smart envi-
           ronments. In: Proceedings of the 7th International Conference on Computer Vision
           Systems (ICVS). Lecture Notes in Computer Science, vol. 5815. Springer, Liège
           (2009)
[SRF10]   Schauerte, B., Richarz, J., Fink, G.A.: Saliency-based identification and recognition of
           pointed-at objects. In: Proceedings of the 23rd International Conference on Intelligent
           Robots and Systems (IROS). IEEE/RSJ, Taipei, Taiwan (2010)
[SRP+09]   Schauerte, B., Richarz, J., Plötz, T., Thurau, C., Fink, G.A.: Multi-modal and multi-
           camera attention in smart environments. In: Proceedings of the 11th International
           Conference on Multimodal Interfaces (ICMI). ACM, Cambridge (2009)
[SS12a]   Schauerte, B., Stiefelhagen, R.: Learning robust color name models from web images.
           In: Proceedings of the 21st International Conference on Pattern Recognition (ICPR).
           IEEE, Tsukuba, Japan (2012)
[SS12b]   Schauerte, B., Stiefelhagen, R.: Predicting human gaze using quaternion DCT image
           signature saliency and face detection. In: Proceedings of the IEEE Workshop on the
           Applications of Computer Vision (WACV). IEEE, Breckenridge, CO, USA (2012)
[SS12c]   Schauerte, B., Stiefelhagen, R.: Quaternion-based spectral saliency detection for eye
           fixation prediction. In: Proceedings of the 12th European Conference on Computer
           Vision (ECCV). Springer, Firenze, Italy (2012)
[SSS14]   Schneider, T., Schauerte, B., Stiefelhagen, R.: Manifold alignment for person inde-
           pendent appearance-based gaze estimation. In: Proceedings of the 21st International
           Conference on Pattern Recognition (ICPR). IEEE, Stockholm, Sweden (2014)
[SWS15a]   Schauerte, B., Wörtwein, T., Stiefelhagen, R.: Color decorrelation helps visual
           saliency detection. In: Proceedings of the 22nd International Conference on Image
           Processing (ICIP). IEEE (2015)
[SZ14]   Schauerte, B., Zamfirescu, C.T.: Small k-pyramids and the complexity of determining
           k. J. Discrete Algorithms (JDA) (2014)
[SLNG07]   Setlur, V., Lechner, T., Nienhaus, M., Gooch, B.: Retargeting images and video for
           preserving information saliency. IEEE Comput. Graph. Appl. **27**(5), 80–88 (2007)
[SI09]   Siagian, C., Itti, L.: Biologically inspired mobile robot vision localization. IEEE
           Trans. Robot. **25**(4), 861–873 (2009)
[SS06]   Simion, C., Shimojo, S.: Early interactions between orienting, visual sampling and
           decision making in facial preference. Vis. Res. **46**(20), 3331–3335 (2006)
[SMI]   SMIvision: Sensomotoric instruments gmbh. http://www.smivision.com/
[STET01]   Spivey, M.J., Tyler, M.J., Eberhard, K.M., Tanenhaus, M.K.: Linguistically mediated
           visual search. Psychol. Sci. **12**, 282–286 (2001)
[SLBJ03]   Suh, B., Ling, H., Bederson, B.B., Jacobs, D.W.: Automatic thumbnail cropping and
           its effectiveness. In: ACM Symposium on User interface Software and Technology
           (2003)
[SW01]   Sussman, E.S., Winkler, I.: Dynamic sensory updating in the auditory system. Cogn.
           Brain Res. **12**, 431–439 (2001)
[Sus05]   Sussman, E.S.: Integration and segregation in auditory scene analysis. J. Acoust. Soc.
           Am. **117**, 1285–1298 (2005)
[TG88]   Treisman, A.M., Gormican, S.: Feature analysis in early vision: evidence from search
           asymmetries. Psychol. Rev. **95**(1), 15–48 (1988)
[TG80]   Treisman, A.M., Gelade, G.: A feature-integration theory of attention. Cogn. Psychol.
           12(1), 97–136 (1980)
[TTDC06]   Triesch, J., Teuscher, C., Deák, G.O., Carlson, E.: Gaze following: why (not) learn
           it? Dev. Sci. **9**(2), 125–147 (2006)

[Uni]   University of British Columbia: Curious George Project. https://www.cs.ubc.ca/labs/
             lci/curious_george/. Accessed 3 April 2014
[WK06]   Walther, D., Koch, C.: Modeling attention to salient proto-objects. Neural Netw.
             19(9), 1395–1407 (2006)
[WWW04]   Weissman, D.H., Warner, L.M., Woldorff, M.G.: The neural mechanisms for mini-
             mizing cross-modal distraction. J. Neurosci. **24**, 10 941–10 949 (2004)
[Wika]   Wikimedia Common (Blausen.com staff): Blausen gallery 2014, ear anatomy. http://
             commons.wikimedia.org/wiki/File:Blausen_0328_EarAnatomy.png. 23 Feb 2015
             (License CC BY 3.0)
[Wikb]   Wikimedia Common (Blausen.com staff): Blausen gallery 2014, the internal ear.
             http://commons.wikimedia.org/wiki/File:Blausen_0329_EarAnatomy_InternalEar.
             png. 23 Feb 2015 (License CC BY 3.0)
[Wikd]   Wikimedia Common (Oarih): Cochlea-crosssection. http://commons.wikimedia.org/
             wiki/File:Cochlea-crosssection.png. 23 Feb 2015 (License CC BY-SA 3.0)
[WTSH+03]   Winkler, I., Teder-Salejarvi, W.A., Horvath, J., Naatanen, R., Sussman, E.: Human
             auditory cortex tracks task-irrelevant sound sources. Neuroreport **14**, 2053–2056
             (2003)
[WCS+05]   Winkler, I., Czigler, I., Sussman, E., Horvath, J., Balazs, L.: Preattentive binding of
             auditory and visual stimulus features. J. Cogn. Neurosci. **17**, 320–339 (2005)
[WS13]   Winkler, S., Subramanian, R.: Overview of eye tracking datasets. In: International
             Workshop on Quality of Multimedia Experience (2013)
[WCS+15]   Woertwein, T., Chollet, M., Schauerte, B., Stiefelhagen, R., Morency, L.-P., Scherer,
             S.: Multimodal public speaking performance assessment. In: Proceedings of the 17th
             International Conference on Multimodal Interaction (ICMI). ACM (2015)
[WSMS15]   Woertwein, T., Schauerte, B., Mueller, K., Stiefelhagen, R.: Interactive web-based
             image sonification for the blind. In: Proceedings of the 17th International Conference
             on Multimodal Interaction (ICMI). ACM (2015)
[WHK+04]   Wolfe, J.M., Horowitz, T.S., Kenner, N., Hyle, M., Vasan, N.: How fast can you
             change your mind? the speed of top-down guidance in visual search. Vis. Res. **44**,
             1411–1426 (2004)
[Wol94]   Wolfe, J.M.: Guided search 2.0: a revised model of visual search. Psychon. Bull. Rev.
             1, 202–238 (1994)
[WCF89]   Wolfe, J.M., Cave, K., Franzel, S.: Guided search: an alternative to the feature integra-
             tion model for visual search. J. Exp. Psychol.: Hum. Percept. Perform. **15**, 419–433
             (1989)
[WBB+96]   Woodruff, P.W., Benson, R.R., Bandettini, P.A., Kwong, K.K., Howard, R.J.,
             Talavage, T., Belliveau, J., Rosen, B.R.: Modulation of auditory and visual cortex
             by selective attention is modality-dependent. Neuroreport **7**, 1909–1913 (1996)
[XZKB10]   Xu, T., Zhang, T., Kühnlenz, K., Buss, M.: Attentional object detection of an active
             multi-vocal vision system. Int. J. Humanoid. **7**(2) (2010)
[Yar67]   Yarbus, A.L.: Eye Movements and Vision. Plenum Press (1967)
[ZTM+08]   Zhang, L., Tong, M.H., Marks, T.K., Shan, H., Cottrell, G.W.: Sun: a bayesian frame-
             work for saliency using natural statistics. J. Vis. **8**(7) (2008)
[GZMT12]   Zhang, L., Tong, M.H., Marks, T.K., Shan, H., Cottrell, G.W.: Context-aware saliency
             detection. IEEE Trans. Pattern Anal. Mach, Intell (2012)

Chapter 3
Bottom-Up Audio-Visual Attention for Scene Exploration

We can differentiate between two attentional mechanisms: First, overt attention directs the sense organs toward salient stimuli to optimize the perception quality. For example, it controls human eye movements in order to project objects of interest onto the fovea of the eye. Second, covert attention focuses the mental processing (e.g., object recognition) of sensory information on the salient stimuli. This is necessary to achieve a high reactivity despite the brain's limited computational resources that are otherwise unable to process the full amount of incoming sensory information. And, it has been formally shown that this mechanism is an essential aspect of human perception, because it transforms several NP-complete perceptual tasks (e.g., bottom-up perceptual search) into computationally tractable problems [Tso89, Tso95].

Since robots have to deal with limited computational resources and the fact that its sensor orientation influences the quality of incoming sensory information, biologically-inspired models of attention have attracted an increasing interest in the field of robotics to optimize the use of resources and improve perception in complex environments. In this chapter, we present how we implemented bottom-up audio-visual overt attention on the head of KIT's robot platform ARMAR [ARA+06, AWA+08]. Our work ranges from developing novel auditory and improved visual saliency models over defining a modality independent 3-dimensional (3D) saliency representation to actually planning in which order the robot should attend salient stimuli. All our methods have in common that they are designed with computational efficiency in mind, because the robot should be able to quickly react to changes in its environment and—in general—a sensible attentional mechanism should not require more resources than it can save.

We first have to define what kind of signals should attract the robot's attention, i.e. what is perceptually "salient". For this purpose, we developed novel auditory and visual saliency models: Our auditory saliency model is based on Itti and Baldi's surprise model [IB06]. According to this model, auditory stimuli that are unexpected given the prior frequency distribution are defined as being salient. This model has a biological foundation, because the spectrogram is similar in function to the human

© Springer International Publishing Switzerland 2016
B. Schauerte, *Multimodal Computational Attention for Scene Understanding and Robotics*, Cognitive Systems Monographs 30,
DOI 10.1007/978-3-319-33796-8_3

Basilar membrane [SIK11] while surprise is related to early sensory neurons [IB06]. Our visual saliency model also relies on the frequency spectrum and suppresses the image's amplitude components to highlight salient image regions. This so-called spectral whitening [OL81] accentuates image regions that depict edges and narrow events that stand out from their surround, which can be related back to Treisman's feature integration theory principles [TG80]. To improve the performance of visual saliency models, we propose to decorrelate the image's color information. This method not just improves the performance of our algorithms, but also the performance of several other visual saliency algorithms as we have shown on three datasets with respect to three evaluation measures. Furthermore, since it has been suggested that there exists a bottom-up attention mechanism for faces [CFK08], we integrate face detection as a bottom-up visual saliency cue.

To model auditory and visually salient stimuli in a common representation, we build upon the notion of salient proto-objects [WK06]—a concept related to Treisman's object files, see Chap. 2—and derive a 3D parametric Gaussian proto-object model. Here, each salient proto-object encodes the perceptual saliency as well as the location and its rough extent of an area in space that is likely to contain an object of interest. To detect and extract salient visual proto-objects, we analyze the saliency map's isophote curvature to extract salient peaks and their contributing pixels, which can then be used to fit a Gaussian model. Acoustic sound source localization is used to locate salient auditory stimuli and the localization uncertainty is encoded as the proto-object's spatial extent, i.e. spatial variance. Since all auditory and visual stimuli are represented by 3D Gaussian weight functions, we are able to efficiently perform crossmodal clustering and proto-object fusion over space and time. The information contained in each cluster is then fused to implement a biologically-plausible crossmodal integration scheme [OLK07].

Given the salient audio-visual proto-objects, we are able use the encoded information about perceptual saliency and location to plan in which order the robot should attend and analyze the proto-objects. We implemented and compared three strategies: Attending the proto-objects in an order that minimizes ego-motion, attending the proto-objects in the order of decreasing saliency, and performing multiobjective optimization to find a trade-off that suits both criteria, i.e. minimize ego-motion while giving priority to highly salient regions.

We demonstrate the applicability of our system and its components in a series of quantitative and qualitative experiments. First, we test the performance of the proposed auditory and visual saliency models. Since our goal was to implement overt attention on a robotic platform, we validate our visual saliency model on human eye tracking data. We show that our approach to visual saliency is state-of-the-art in predicting where humans look in images. Since we could not rely on data analog to eye tracking data to validate our auditory saliency model, we follow a more practice-oriented approach and show that our model is able to reliably detect arbitrary salient auditory events. To validate the overall system behavior, we first performed a series of qualitative active perception experiments. Second, we demonstrate that using multiobjective optimization we can effectively reduce the necessary ego-motion while still assigning a high priority to more salient proto-objects, which results in more efficient scan path patterns.

Remainder

Complementary to our broad background presentation in Chap. 2, we provide a detailed overview of related work (Sect. 3.1) that is relevant to understand the contributions presented in this chapter. Then, we present and evaluate our visual (Sect. 3.2) and auditory (Sect. 3.3) saliency model. Subsequently, we describe our audio-visual saliency-driven scene exploration system (Sects. 3.4 and 3.5). We discuss how we map the detected salient acoustic events and visually salient regions in a common, modality-independent proto-object representation. Then, we describe how the common salient proto-object representation allows us to fuse this information across modalities. Afterwards, we explain how we can plan the robot's eye movement based on our audio-visually salient proto-objects.

3.1 Related Work and Contributions

In the following, we first present the state-of-the-art—excluding our work that is presented in this book—for each of the affected research topics. Then, after each topic's overview, we discuss our contribution with respect to the state-of-the-art.

3.1.1 Spectral Visual Saliency

The first spectral approach for visual saliency detection was presented in 2007 by Hou et al. [HZ07]. Since then, several spectral saliency models have been proposed (see, e.g., [BZ09, GZ10, GMZ08, PI08b, HZ07, AS10, LLAH11]). Hou et al. proposed to use the Fourier transform to calculate the visual saliency of an image. To this end,—processing each color channel separately—the image is Fourier transformed and the magnitude components are attenuated. Then, the inverse Fourier transform is calculated using the manipulated magnitude components in combination with the original phase angles. The saliency map is obtained by calculating the absolute value of each pixel of this inverse transformed image and subsequent Gaussian smoothing. This way Hou et al. achieved state-of-the-art performance for salient region (proto-object) detection and psychological test patterns. However, although Hou et al. were the first to propose this method for saliency detection, it has been known for at least three decades that suppressing the magnitude components in the frequency domain highlights signal components such as lines, edges, or narrow events (see [OL81, HBD75]).

In 2008 [PI08b], Peters et al. analyzed the role of Fourier phase information in predicting visual saliency. They extended the model of Hou et al. by linearly combining the saliency of the image at several scales. Then, they analyzed how well this model predicts eye fixations and found that "salience maps from this model significantly predicted the free-viewing gaze patterns of four observers for 337 images of natural outdoor scenes, fractals, and aerial imagery" [PI08b].

Also in 2008 [GMZ08], Guo et al. proposed the use of quaternions as a holistic color image representation for spectral saliency calculation. This was possible because quaternions provide a powerful algebra that allows to realize a hypercomplex Fourier transform [Ell93], which was first demonstrated to be applicable for color image processing by Sangwine [San96, SE00]. Thus, Guo et al. were able to Fourier transform the image as a whole and did not have to process each color channel separately. Furthermore, this made it possible to use the scalar part of the quaternion image as 4th channel to integrate a motion component. However, in contrast to Hou et al., Guo et al. did not preserve any magnitude information and perform a whitening instead. Most interestingly, Guo et al. were able to determine salient people in videos and outperformed the models of Itti et al. [IKN98] and Walther et al. [WK06]. In 2010, a multiresolution attention selection mechanism was introduced, but the definition of the main saliency model remained unchanged [GZ10]. However, most interestingly, further experiments demonstrated that the approach outperformed several established approaches in predicting eye gaze on still images.

In 2009 [BZ09], Bian et al. adapted the work of Guo et al. by weighting the quaternion components. Furthermore, they provide a biological justification for spectral visual saliency models and—without any detailed explanation—proposed the use of the YUV color space, in contrast to the use of the previously applied intensity and color opponents (ICOPP) [GMZ08, GZ10], and RGB [HZ07]. This made it possible to outperform the models of Bruce et al. [BT09], Gao et al. [GMV08], Walther and Koch [WK06], and Itti and Koch [IKN98] when predicting human eye fixations on video sequences.

In 2012 [HHK12], Hou et al. proposed and theoretically analyzed the use of the discrete cosine transform (DCT) for spectral saliency detection. They showed that this approach outperforms all other evaluated approaches—including the algorithms of Itti and Koch [IKN98], Bruce and Tsotsos [BT09], Harel et al. [HKP07], and Zhang et al. [ZTM+08]—in predicting human eye fixations on the well-known Toronto dataset [BT09]. Furthermore, Hou et al. pointed out the importance of choosing an appropriate color space.

Contributions

We combine and extend several aspects of spectral saliency detection algorithms. Analog to Guo et al.'s [GMZ08] adaptation of Hou et al.'s spectral residual saliency algorithm [HZ07], we extend Hou et al.'s DCT image signature approach [HHK12] and use quaternions to represent and process color images in a holistic framework. Consequently, we apply the quaternion discrete cosine transform (QDCT) and signum function to calculate the visual saliency. Furthermore, we integrate and investigate the influence of quaternion component weights as proposed by Bian et al. [BZ09], adapt the multiscale model by Peters et al. [PI08b], and propose the use of the quaternion eigenaxis and eigenangle for saliency algorithms that rely on the quaternion Fourier transform (e.g., [HZ07, GZ10, GMZ08]). This way, we were able to improve the state-of-the-art in predicting where humans look on three eye tracking datasets—proving the outstanding performance of spectral models for this task, which was not conclusively shown before.

3.1.2 Visual Saliency and Color Spaces

As has been noted for spectral saliency models (see Sect. 3.1.1), the chosen base color space can have a significant influence on the performance of bottom-up visual saliency models. Accordingly, different color spaces have been used for saliency models (see also Sect. 3.1.1) such as, for example, RGB (e.g., [HZ07]), CIE Lab (e.g., [HHK12]), and ICOPP (e.g., [GMZ08, GZ10]). The most prominent color space is probably red-green-blue (RGB), which is an additive color space that is suited for most of today's displays (see [Pas08]). The YUV model defines a color space in terms of one luma (Y) and two chrominance (UV) components. Similarly, the CIE 1976 Lab color space has been designed to approximate human vision and aspires to perceptual uniformity. The simple intensity and red-green/blue-yellow color opponent (ICOPP) model is often used in conjunction with saliency models (e.g., [GMZ08]). The LMS color space models the response of the three types of cones of the human eye, which are named after their sensitivity for long, medium and short wavelengths [SG31]. Whereas Geusebroek et al.'s Gaussian color space [GvdBSG01] represents an extension of the differential geometry framework into the spatio-spectral domain.

Decorrelation of color information has been successfully applied for several applications, e.g. texture analysis and synthesis [LSL00, HB95], color enhancement [GKW87], and color transfer [RP11]. More importantly, it is highly related to the human visual system and techniques such as the zero-phase transform (ZCA) have been developed and proposed to model aspects of the human visual system [BS97]. Buchsbaum and Gottschalk [BG83] and Ruderman et al. [RCC98] found that linear decorrelation of LMS cone responses at a point matches the opponent color coding in the human visual system. However, when modeling the human visual system it is mostly applied in the context of spatio-chromatic decorrelation, i.e. local (center-surround) contrast filter operations [BSF11, RCC98, BS97]. Since decorrelation is an important aspect of the human visual system, it has also been part of a few visual saliency models. Duan et al. [DWM+11] explored the use of principal component analysis (PCA) on image patches, which is closely related Zhou et al.'s approach [ZJY12] in which the image is first segmented into patches and then the PCA is used to reduce the patch dimensions to throw out dimensions that are basically noise for the saliency calculation. Similarly, Wu et al. [WCD+13] propose to use the PCA to attenuate noise as well as to reduce computational complexity. Luo et al. [LLLN12] also use the PCA on a block-wise level to differentiate between salient objects and background.

Contributions

As a result of our experience with quaternion-based spectral algorithms, we wanted to try the opposite approach: Instead of using quaternions to represent and process the image's color information holistically, we try to decorrelate the information in the color components. To this end, we propose to use a global image-dependent decorrelated color space for visual saliency detection. This way, we are able to improve the performance of all eight visual saliency algorithms that we tested: Itti and Koch's classic model [IKN98], Harel's graph-based visual saliency [HKP07], Hou and Zhang's pure Fourier transform algorithm [HZ07], Hou et al.'s DCT image signature [HHK12], Lu and Lim's histogram-based approach [LL12], Achanta's frequency-tuned approach [AHES09], and our own QDCT image signature and quaternion Fourier transform with eigenaxis/eigenangle algorithms.

3.1.3 Visual Saliency and Faces

Studies have shown that—independent of the subject's task—when looking at natural images the gaze of observers is attracted to faces (see [CFK08, SS06]). Even more, there exists evidence that the gaze of infants is attracted by face-like patterns before they can consciously perceive the category of faces [SS06], which is supported by studies of infants as young as 6 weeks that sugggest that faces are visually captivating [CC03]. This seems to play a crucial role in early development, especially emotion and social processing (see, e.g., [KJS+02]). This early attraction and inability to avoid looking at face-like patterns suggests that there exist bottom-up attention mechanisms for faces [CFK08]. To model this influence, Cerf et al. combined traditional visual saliency models—Harel's graph-based visual saliency (GBVS) [HKP07] and Itti and Koch's model [IKN98]—with face detections provided by the well-known Viola-Jones detector [CHEK07, CFK09].

Contributions

We build on Cerf et al.'s work and integrate a scalable Gaussian face model based on modified census transform (MCT) face detection [FE04] into our state-of-the-art low-level visual saliency model. This way, we are able to improve the state-of-the-art in predicting where people look in the presence of faces. Furthermore, considering the face detections and bottom-up visual saliency as two modalities, we investigate the influence of different biologically plausible combination schemes (see [OLK07]).

3.1.4 Auditory Saliency

As has already been addressed in Sect. 2.1.2, in contrast to the vast amount of proposed visual saliency models (cf. [FRC10, Tso11]), only few computational bottom-up auditory attention models exist. Most closely related to our work is the model described by Kayser et al. [KPLL05], see Fig. 2.8 on p. 37, which has later been adopted by Kalinli and Narayanan [KN09]. This model is based on the well-established visual saliency model of Itti and Koch [IKN98] and, most notably, has been successfully applied to speech processing by Kalinli et al. [KN09] and, in principle, by Lin et al. [LZG+12] to allow for faster human acoustic event detection through audio visualization.

Contributions

The application of Itti and Koch's visual saliency model to spectrograms has several drawbacks: First and most importantly, it requires that the spectrogram has elements of the future to detect salient events in the present, which prohibits online detection and—as a consequence—quick reactions to salient acoustic events. This is caused by the inherent down-scaling and filtering in Itti and Koch's model, which makes precise localization of salient stimuli at the borders problematic. Second, Itti and Koch's model is computationally expensive, because it requires the calculation and combination of a considerable amount of 2D feature maps at each time step. Third, although Itti and Koch's saliency model represents an outstanding historical accomplishment, it can hardly be said to be state-of-the-art [BSI13b]. To account for these drawbacks, we developed auditory Bayesian surprise, see Sect. 3.3.

3.1.5 Audio-Visual Saliency-Based Exploration

To realize overt audio-visual attention, it is not sufficient to just determine auditory or visually salient stimuli, but is is also necessary to meaningfully and efficiently integrate the information from both modalities. This is a topic that seems to attract increasing attention, however only a relatively modest number of theoretical studies (e.g., [OLK07]) and models have been proposed so far (e.g., [RMDB+13, SPG12]) The proposed models rely on the existence of a 2-dimensional (2D) audio-visual saliency map [RMDB+13, OLK07], although it is unclear whether a similar representation exists in the human brain and how such a 2D spatial auditory saliency map can be calculated. Furthermore, it is also unclear how such a map could be updated in the presence of ego-motion. However, when realizing overt attention it is important to consider that each shift of the overt focus of attention leads to ego-motion, which partially renders the previously calculated information obsolete [BKMG10]. Accordingly, it is necessary to enable storing and updating the saliency as well as object information in the presence of ego-motion that are caused by overt attention shifts.

Saliency-based overt attention, i.e. directing the robot sensors toward salient stimuli, and saliency-based scene exploration has been addressed by several authors in recent years (e.g. [MFL+07, BKMG10, RLB+08, XCKB09, VCSS01, FPB06, DBZ07, YGMG13]). Almost all state-of-the-art systems only consider visual attention (e.g. [MFL+07, BKMG10, DBZ07, OMS08]), which—among other drawbacks—makes it impossible to react on salient events outside the visual field of view (cf.[SRP+09]). Most related to our work on audio-visual attention are the approaches by Ruesch et al. [RLB+08], who implement audio-visual attention for the "iCub" [Rob] robot platform, and Schauerte et al. [SRP+09], who implement an audio-visually attentive smart room. Both systems use common visual saliency algorithms (see Sect. 2.1.1) and the energy of the audio signal as primitive auditory attention model, due to the absence of applicable, more elaborate auditory saliency models. Ruesch et al. [RLB+08] use a linear combination of audio-visual stimuli, whereas Schauerte et al. [SRP+09] use Fuzzy logic [Zad65] to implement a divers set of audio-visual combinations, including linear combinations. More importantly, Ruesch et al. [RLB+08] rely on an ego-centric spatial grid representation, i.e. azimuth-elevation maps, following the idea of a sensory ego-sphere in which the reference coordinate system is anchored to a fixed point on the robot's body [FPB06]. Schauerte et al. [SRP+09] use a 3D voxel representation, which is similar to Meger et al.'s 2D occupancy grid representation [MFL+07]. Meger et al.'s [MFL+07] and Schauerte et al.'s [SRP+09] representation anchor the reference coordinate system to a fixed point in the scene to allow for ego-motion.

In most publications on overt attention, the order in which the objects in the scene are attended is solely based on the perceptual saliency (see, e.g., [RLB+08, BZCM08, IK00]; [SRP+09]). Accordingly, in each focus of attention selection step, the location with the highest saliency gains the focus of attention and an inhibition of return mechanism ensures that salient regions are not visited twice. However, in many practical applications, it is beneficial to incorporate other aspects into the decision which location to attend next; for example, sensor coverage planning to maximize the coverage of previously unseen areas [MFL+07], top-down target information for visual search [OMS08, WAD09, Wel11], transsaccadic memory consistency [WAD11], or a task-dependent spatial bias [DBZ07]. Most related to our work, Saidi et al. [SSYK07] and Andreopoulos et al. [AHW+10] use rating functions for object search such as, most importantly, a motion cost function for sensor alignment.

Contributions

Not unlike Kahneman and Treisman's object files [KTG92], in coherence theory of visual cognition, proto-objects are volatile units of information that can be accessed by selective attention and subsequently validated as actual objects [WK06]. Our audio-visual saliency representation relies on the concept of audio-visual proto-objects, since we propose a parametric object-centred crossmodal 3D model that is based on Gaussian weight functions to represent salient proto-object regions. For this purpose, we use the visual saliency maps' isophote curvature and stereo vision (see [LHR05]) to extract visual proto-object regions and use sound source localization and salient acoustic event detection to model auditory salient proto-objects. This proto-object model allows for efficient representation, fusion, and update of information. By treating the proto-object regions as primitive, uncategorized object entitities in our world model, it is also the foundation to implement our exploration strategies and object-based inhibition of return. Apart from this seamless model integration, our parametric representation has further practical advantages compared to grid representations. Most importantly, every spatial grid representation leads to a spatial quantization and consequently localization error. To reduce this error and increase the model's quality, it is necessary to increase the grid resolution which typically has a quadratic or cubic impact on the run-time of algorithms that operate on 2D pixels or 3D voxels, respectively. In contrast, our model does not have a quantization error and the run-time of, for example, crossmodal saliency fusion only depends on the number of salient proto-objects, which according to the definition of salient signals is relatively small.

To plan which location to attend next, we use a flexible, multiobjective exploration strategy based on the salient proto-object regions. In our current implementation, our multiobjective target function considers two criteria: the audio-visual saliency and the required head ego-motion. This way, we are able to implement a solution that substantially reduces head ego-motion while it still strongly favors to attend highly salient regions as fast as possible. Furthermore, the chosen formulation makes it easily possible to integrate additional target criteria, e.g. task-specific influences, in the future. Similar to human behavior (cf. [Hen03, Wel11]), our exploration considers all salient regions that are present in the short-time memory of our world model, even if they are currently outside the robot's view. An integrated tracking of proto-objects—which can be linked to already attended objects—makes it possible to detect changes in object saliency and to distinguish novel proto-objects from already attended objects. This way, this makes it possible to seamlessly implement object-based inhibition of return (IoR), which is consistent with human behavior [TDW91] and has the advantage that we are able to realize IoR even for moving targets.

3.1.6 Scene Analysis

Fusing the information of different sensors and sensor modalities in order to analyze a scene has been addressed throughout the years in several application areas (see, e.g., [Ess00, MSKK10, KBS+10, HY09, HL08]). We build on Machmer et al.'s hierarchical, knowledge-driven audio-visual scene analysis approach [MSKK10] that follows an integrated bottom-up and top-down analysis methodology (see [HL08, HY09]). In this framework, the multimodal classification and fusion at each level of the knowledge hierarchy is done bottom-up whereas the appropriate selection of classification algorithms is done in a top-down fashion. The basis for this exploration and analysis is an object-based world model as proposed by Kühn et al. [KBS+10], which provides an uncertainty-based description for every object attribute. A notable feature of the chosen object analysis approach is that it facilitates the dynamic

Fig. 3.1 An example to illustrate the principle of the hierarchical object analysis. "*Blue*" attributes trigger a refinement in the hierarchy and "*green*" attributes supply additional information about an entity

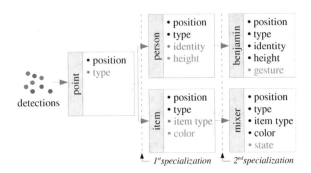

adjustment of object-specific tracking parameters, e.g. for mean shift [CM02], depending on the classification result, e.g. person or object specific parameters.

Contributions

We integrated saliency-driven, iterative scene exploration into a hierarchical, knowledge-driven audio-visual scene analysis approach that was first presented by Machmer et al. [MSKK10], see Fig. 3.1. This, in principle, consistently implements many of the ideas expressed in Treisman's psychological attention model, see Fig. 2.1.

3.2   Visual Attention

Various saliency models have been proposed (see Sects. 2.1.1 and 3.1.1) that vary substantially in what they highlight as being "salient", i.e. what should or is most likely to attract the attention. However, in principle, all saliency models share the same principles and target description, i.e. use a set of image features and contrast measures to highlight "sparse" image regions. Such sparse image regions contain or consist of rare features or other irregularities that let them visually "pop out" from the surrounding image context.

In this context, the most fundamental features are color features and edge orientations (see Figs. 2.1, 2.3 and 2.4). Here, the opponent color theory [Her64] forms the theoretical justification for the widely applied opponent color model. In this model, it is suggested that the human visual system interprets information about color based on three opponent channels: red versus green, blue versus yellow, and black versus white, see Fig. 3.2. The latter is achromatic and consequently encodes luminance while the other components encode chrominance. The color opponents can be seen to represent nearly orthogonal axes in an image-independent color space in which red/green, blue/yellow, and white/black form the start/end points of each axis, because—under normal viewing conditions—there exists no hue that humans could describe as a mixture of opponent hues. For example, there exists no hue that appears at the same time read and green (i.e., "redgreen'ish") or yellow and blue (i.e., "yellowblue'ish") to a human observer. Compression and efficient coding of

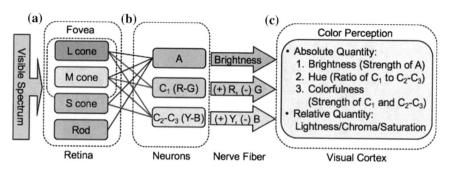

Fig. 3.2 The opponent color model as input to the visual cortex. Image from Wikimedia [Wikc]

sensory signals is another approach to address this aspect and in this context it was found that decorrelation of LMS cone responses at a point matches the opponent color coding in the human visual system [BG83, RCC98]. We further investigate this topic and will first witness the influence of color space on spectral saliency models (Sect. 3.2.1) before we investigate color space decorrelation as a preprocessing or feature encoding step for visual saliency detection (Sect. 3.2.2).

To determine what "pops out" of the feature maps, we rely on spectral visual saliency models. Such models were first proposed to detect proto-objects in images [HZ07] and subsequently it was shown that spectral models also provide an outstanding performance in predicting where people look (e.g., [BZ09, HHK12]). However, real-valued spectral saliency models have—like many other methods—the problem that they process the color channels independently, which can lead to a loss or misrepresentation of information that can result in a suboptimal performance (see, e.g., Fig. 3.3). As an alternative, it is possible to represent images as quaternion matrices and use the quaternion algebra to process the color information as a whole (e.g., [ES07, SE00, BZ09]), i.e. holistically. We present how we can calculate the visual saliency based on the quaternion discrete cosine transform and, since not all components might have the same importance for visual saliency, that weighting the quaternion components can improve the performance.

Human attention is not just sensitive to low-level bottom-up features such as, most importantly, color and intensity contrast. Instead, there exists evidence that some complex cues can as well attract human attention independent of task, which suggests that such cues are bottom-up and not top-down features. Most importantly, it has been shown that faces and face-like patterns attract human attention [CFK08]. Consequently, in addition to color features, we integrate face detection as an attentional bottom-up cue in our model (Sect. 3.2.3).

Remainder

The remainder of this section is organized as follows: First, we present real-valued and quaternion-based spectral saliency detection (Sect. 3.2.1). Then, we introduce color space decorrelation to boost the performance of several visual saliency algorithms (Sect. 3.2.2). Finally, we show how we integrate the influence of faces into our visual attention model (Sect. 3.2.3).

(a)   **(b)**   **(c)**

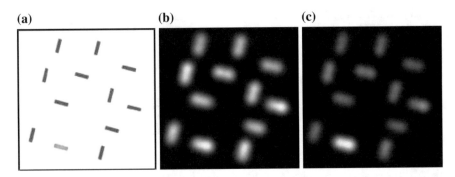

Fig. 3.3 A simple, illustrated example of the disadvantage of processing an image's color channels separately. Image from [BZ09] reprinted with permission from Springer. **a** Psychological popout color pattern. **b** Saliency map generated by computing each color channel separately. **c** Saliency map generated by considering the image as a quaternion matrix

3.2.1 Spectral Visual Saliency

Hou et al. introduced the spectral residual saliency model to detect salient proto-objects in images [HZ07], which also form a foundation for our audio-visual exploration system as will be addressed in Sect. 3.4. Spectral saliency is based on the idea that "statistical singularities in the spectrum may be responsible for anomalous regions in the image, where proto-objects pop up" [HZ07]. To detect salient image regions, Hou et al. [HZ07] attenuate the magnitude in the Fourier frequency spectrum, which in its extreme form leads to a phase-only reconstruction of the image—i.e., the phase-only Fourier transformed signal with unity magnitude—which is known as spectral whitening [OL81]. Although the application of this idea to visual saliency detection was novel, the principle has been widely known for a long time in signal processing theory. As Oppenheim described it [OL81], "since the spectral magnitude of speech and pictures tends to fall off at high frequencies, the phase-only signal $f_p(x)$[1] will, among other effects, experience a high-frequency emphasis which will accentuate lines, edges and other narrow events without modifying their position". Back in 1981, computational visual saliency has not been an active research field and accordingly Oppenheim focused on other applications such as, e.g., image coding and reconstruction. However, since the basic principle of visual saliency models is to highlight such edges and sparse, narrow image regions, what Oppenheim described in 1981 was a visual saliency model that should become the state-of-the-art 25 years later.

It is possible to calculate the spectral saliency based on the Fourier transform [HZ07] as well as on the cosine transform [HHK12]. Unfortunately, if we want to calculate the spectral saliency for color images, it is necessary to process each image channel separately and subsequently fuse the information. However, since the color

---

[1]Oppenheim refers to $\mathscr{F}[f_p(x)] = \frac{1}{|F(\omega)|}\mathscr{F}[f(x)]$ with $F(\omega) = \mathscr{F}[f](\omega)$.

space components and consequently the information across the image channels is not independent, this means that color information is involuntarily mishandled or even lost, see Fig. 3.3. An interesting development with regard to this problem is the use of quaternions as a holistic representation to process color images [ES07, GMZ08, BZ09]. The quaternion algebra makes it possible to process color images as a whole without the need to process the image channels separately and, in consequence, tear apart the color information. Interestingly, since the Fourier transform and the cosine transform are also well-defined in the quaternion algebra, we can holistically calculate the spectral saliency based on quaternion color images.

3.2.1.1 Real-Valued Spectral Saliency

Spectral Residual and Whitening

Given a single-channel image I_1, we can calculate the phase angle P and amplitude A of the image's Fourier frequency spectrum

$$P = \Phi(\mathscr{F}(I_1)) \tag{3.1}$$
$$A = |\mathscr{F}(I_1)|. \tag{3.2}$$

The spectral residual saliency map S_{FFT} [HZ07] of the image can then be calculated according to

$$S_{\text{FFT}} = \mathscr{S}_{\text{FFT}}(I_1) = g * |\mathscr{F}^{-1}\{e^{R+iP}\}| \quad \text{with} \tag{3.3}$$
$$L(x, y) = \log A(x, y) \quad \text{and} \tag{3.4}$$
$$R(x, y) = L(x, y) - [h * L](x, y). \tag{3.5}$$

Here, $\mathscr{F}$ denotes the Fourier transform; g and h are Gaussian filter kernels. h is applied to substract the smoothed log magnitude, i.e. $h * L$, from the raw log magnitude L, which forms the "spectral residual" R. In principle, this process implements a local contrast operation in the log magnitude matrix, whose strength is defined by the variance σ_h of the Gaussian filter h.

Shortly after Hou et al.'s method was proposed, Guo et al. showed [GMZ08] that the influence of the spectral residual itself is negligible in many situations. This means that R in Eq. 3.3 can be removed

$$S_{\text{PFT}} = \mathscr{S}_{\text{PFT}}(I) = g * |\mathscr{F}^{-1}\{e^{iP}\}|, \tag{3.6}$$

which leads to spectral whitening and is commonly referred to as "pure Fourier transform". However, spectral whitening can be seen as an extreme case of the spectral residual, because the spectral residual R approaches 0 when σ_h approaches 0, i.e. $\lim_{\sigma_h \to 0^+} R = 0$

If we want to process multi-channel color images I, it is necessary to calculate the saliency of each image channel I_c and subsequently fuse the maps, because the 2D Fourier transform is only defined for single-channel images. Consequently, the real-valued spectral saliency for color channel images is defined as [HHK12]

$$S_{\text{FFT}}^C = \mathscr{S}_{\text{FFT}}^C(I) = g * \sum_{1 \le c \le C} \mathscr{S}_{\text{FFT}}(I_c). \tag{3.7}$$

DCT Image Signature

The visual saliency based on discrete cosine transform (DCT) image signatures S_{DCT} for a multi-channel image I is defined as follows [HHK12]:

$$S_{\text{DCT}}^C = \mathscr{S}_{\text{DCT}}^C(I) = g * \sum_{1 \le c \le C} [T(I_c) \circ T(I_c)] \quad \text{with} \tag{3.8}$$

$$T(I_c) = \mathscr{D}^{-1}(\text{sgn}(\mathscr{D}(I_c))), \tag{3.9}$$

where I_c is the cth image channel, $\circ$ denotes the Hadamard—i.e., element-wise—product, sgn is the signum function, $\mathscr{D}$ denotes the DCT, and g is typically a Gaussian smoothing filter. Most notably, it has been formally shown that the DCT image signatures, i.e. $\text{sgn}(\mathscr{D}(I_c))$, suppress the background and are likely to highlight sparse salient features and objects [HHK12].

3.2.1.2 Quaternion Image Processing

Quaternions form a 4-dimensional (4D) algebra $\mathbb{H}$ over the real numbers and are in principle an extension of the 2D complex numbers [Ham66]. A quaternion q is defined as $q = a + bi + cj + dk \in \mathbb{H}$ with $a, b, c, d \in \mathbb{R}$, where i, j, and k provide the basis to define the (Hamilton) product of two quaternions q_1 and q_2 ($q_1, q_2 \in \mathbb{H}$):

$$q_1 q_2 = (a_1 + b_1 i + c_1 j + d_1 k)(a_2 + b_2 i + c_2 j + d_2 k), \tag{3.10}$$

where $i^2 = j^2 = k^2 = ijk = -1$. Since, for example, by definition $ij = k$ while $ji = -k$ the Hamilton product is not commutative. Accordingly, we have to distinguish between left-sided and right-sided multiplications (marked by L and R, respectively, in the following). A quaternion q is called real, if $x = a + 0i + 0j + 0k$, and pure (imaginary), if $q = 0 + bi + cj + dk$. We can define the operators $S(q) = a$ and $V(q) = bi + cj + dk$ that extract the scalar part and the imaginary part of a quaternion $q = a + bi + cj + dk$, respectively. As for complex numbers, we can define conjugate quaternions $\bar{q}$

$$\bar{q} = a - bi - cj - dk \tag{3.11}$$

as well as the norm $|q|$

$$|q| = \sqrt{q \cdot \bar{q}}. \tag{3.12}$$

Here, a unit quaternion is defined as being a quaternion of norm one. Furthermore, we can define the quaternion scalar product $* : \mathbb{H} \times \mathbb{H} \to \mathbb{R}$

$$s = q_1 * q_2 = a_1 a_2 + b_1 b_2 + c_1 c_2 + d_1 d_2. \tag{3.13}$$

Eigenaxis and Eigenangle

Euler's formula for the polar representation using the complex exponential generalizes to a (hypercomplex) quaternion form

$$e^{\mu \Phi} = \cos \Phi + \mu \sin \Phi, \tag{3.14}$$

where μ is a unit pure quaternion (see [SE00] and [GZ10]). Consequently, any quaternion q may be represented in a polar representation such as:

$$q = |q| e^{\gamma \Phi} \tag{3.15}$$

with the norm $|q|$, its "eigenaxis" γ

$$\gamma = f_\gamma(q) = \frac{V(q)}{|V(q)|}, \tag{3.16}$$

and the corresponding "eigenangle" Φ

$$\Phi = f_\Phi(q) = \arctan\left(\frac{|V(q)| \mathrm{sgn}(V(q) * \gamma)}{S(q)}\right) \tag{3.17}$$

with respect to the eigenaxis γ, which is a unit pure quaternion, and where $\mathrm{sgn}(\cdot)$ is the signum function (see [SE00]). The eigenaxis γ specifies the quaternion direction in the 3D space of the imaginary, vector part and can be seen as being a generalization of the imaginary unit of complex numbers. Analogously, the eigenangle Φ corresponds to the argument of a complex number.

Quaternion Images

Every image $\mathbf{I} \in \mathbb{R}^{M \times N \times C}$ with at most 4 color components, i.e. $C \leq 4$, can be represented using a $M \times N$ quaternion matrix

$$\mathbf{I_Q} = I_4 + I_1 i + I_2 j + I_3 k \tag{3.18}$$

$$= I_4 + I_1 i + (I_2 + I_3 i)j \quad \text{(symplectic form),} \tag{3.19}$$

where $\mathbf{I}_c$ denotes the $M \times N$ matrix of the cth image channel. It is common to represent the (potential) 4th image channel as the scalar part (see, e.g., [SE00]), because when using this definition it is possible to work with pure quaternions for the most common color spaces such as, e.g., RGB, YUV and Lab.

Quaternion Discrete Fourier Transform

We can transform a $M \times N$ quaternion matrix $\mathbf{f}$ using the definition of the quaternion Fourier transform $\mathscr{F}_Q^L$ [ES07]:

$$\mathscr{F}_Q^L[f](u, v) = F_Q^L(u, v) \tag{3.20}$$

$$F_Q^L(u, v) = \frac{1}{\sqrt{MN}} \sum_{m=0}^{M-1} \sum_{n=0}^{N-1} e^{-\eta 2\pi((mv/M)+(nu/N))} f(m, n),$$

see Fig. 3.4 for an example. The corresponding inverse quaternion discrete Fourier transform $\mathscr{F}_Q^{-L}$ is defined as:

$$\mathscr{F}_Q^{-L}[F](m, n) = f_Q^L(m, n) \tag{3.21}$$

$$f_Q^L(m, n) = \frac{1}{\sqrt{MN}} \sum_{u=0}^{M-1} \sum_{v=0}^{N-1} e^{\eta 2\pi((mv/M)+(nu/N))} F(u, v).$$

Fig. 3.4 Visualization [ES07] of the quaternion Fourier spectrum of an example image for two transformation axes (1st and 2nd). Example image from the Bruce/Toronto dataset [BT09]. **a** image, **b** 1st eigenangles, **c** 1st eigenaxes, **d** magnitude (log), **e** 2nd eigenangles, **f** 2nd eigenaxes

Here, η is a unit pure quaternion, i.e. $\eta^2 = -1$, that serves as an axis and determines a direction in the color space. Although the choice of η is arbitrary, it is not without consequence (see [ES07, Sect. V]). For example, in RGB a good axis candidate would be the "gray line" and thus $\eta = (i+j+k)/\sqrt{3}$. In fact, as discussed by Ell and Sangwine [ES07], this would decompose the image into luminance and chrominance components.

Quaternion Discrete Cosine Transform

Following the definition of the quaternion DCT [FH08], we can transform the $M \times N$ quaternion matrix f:

$$\mathcal{D}_Q^L[f](p, q) = \alpha_p^M \alpha_q^N \sum_{m=0}^{M-1} \sum_{n=0}^{N-1} \eta f(m, n) \beta_{p,m}^M \beta_{q,n}^N \qquad (3.22)$$

$$\mathcal{D}_Q^R[f](p, q) = \alpha_p^M \alpha_q^N \sum_{m=0}^{M-1} \sum_{n=0}^{N-1} f(m, n) \beta_{p,m}^M \beta_{q,n}^N \eta, \qquad (3.23)$$

where η is again a unit (pure) quaternion that serves as DCT axis. In accordance with the definition of the traditional type-II DCT, we define α and N as follows[2]:

$$\alpha_p^M = \begin{cases} \sqrt{\frac{1}{M}} & \text{for } p = 0 \\ \sqrt{\frac{2}{M}} & \text{for } p \neq 0 \end{cases} \qquad (3.24)$$

$$\beta_{p,m}^M = \cos\left[\frac{\pi}{M}(m + \frac{1}{2})p\right]. \qquad (3.25)$$

Consequently, the corresponding inverse quaternion DCT is defined as follows:

$$\mathcal{D}_Q^{-L}[F](m, n) = \sum_{p=0}^{M-1} \sum_{q=0}^{N-1} \alpha_p^M \alpha_q^N \eta F(p, q) \beta_{p,q}^M \beta_{m,n}^N \qquad (3.26)$$

$$\mathcal{D}_Q^{-R}[F](m, n) = \sum_{p=0}^{M-1} \sum_{q=0}^{N-1} \alpha_p^M \alpha_q^N F(p, q) \beta_{p,q}^M \beta_{m,n}^N \eta. \qquad (3.27)$$

Again, the choice of the axis η is arbitrary (see [ES07]).

---

[2]From a visual saliency perspective, it is not essential to define the case in α that handles $p = 0$. However, this makes the DCT-II matrix orthogonal, but breaks the direct correspondence with a real-even DFT of half-shifted input. Even more, it is possible to entirely operate without normalization, i.e. remove the α terms, which results in a scale change that is irrelevant for saliency calculation.

As can be seen when comparing Eqs. 3.20 and 3.22, the definition of $\mathscr{D}_Q^L$ is substantially different from $\mathscr{F}_Q^L$, because the factors $\beta_{u,m}^M$ are real-valued instead of the hypercomplex terms of $\mathscr{F}_Q^L$. However, both definitions share the concept of a unit pure quaternion η that serves as a transformation axis.

3.2.1.3 Quaternion-Based Spectral Saliency

Eigenaxis and -Angle Fourier Spectral Saliency

Similar to the real-numbered definition of the spectral residual by Hou et al. [HZ07], let A_Q denote the amplitude, E_γ the eigenaxes, and the eigenangles E_Θ (see Sect. 3.2.1.2) of the quaternion image I_Q:

$$E_\gamma(x, y) = f_\gamma(I_Q(x, y)) \tag{3.28}$$

$$E_\Theta(x, y) = f_\Theta(I_Q(x, y)) \tag{3.29}$$

$$A_Q(x, y) = |I_Q(x, y)|. \tag{3.30}$$

Then, we calculate the log amplitude and a low-pass filtered log amplitude using a Gaussian filter h_{σ_A} with the standard deviation σ_A to obtain the spectral residual R_Q:

$$L_Q(x, y) = \log A_Q(x, y) \tag{3.31}$$

$$R_Q(x, y) = L_Q(x, y) - \left[h_{\sigma_A} * L_Q\right](x, y). \tag{3.32}$$

Finally, we can calculate the Eigen spectral residual (ESR) saliency map S_{ESR} using the spectral residual R_Q, the eigenaxis E_γ, and the eigenangle E_Θ:

$$S_{ESR} = \mathscr{S}_{ESR}(I_Q) = h_{\sigma_S} * |\mathscr{F}_Q^{-L}\left[e^{R_Q + E_\gamma \circ E_\Theta}\right]|, \tag{3.33}$$

where $\circ$ denotes the Hadamard product and h_{σ_S} is a real-valued Gauss filter with standard deviation σ_S. If σ_A approaches zero, then the spectral residual R_Q will become 0, i.e. $\lim_{\sigma_A \to 0^+} R_Q(x, y) = 0$, in which case we refer to the model as the Eigen pure quaternion Fourier transform (EPQFT).

If the input image is a single-channel image, then the quaternion definitions and equations are reduced to their real-valued counterparts, in which case Eq. 3.33 is identical to the single-channel real-numbered definitions by Hou et al. [HZ07] and Guo et al. [GMZ08]. Our ESR and EPQFT definition that is presented in Eq. 3.33 differs from Guo's pure quaternion Fourier transform (PQFT) [GMZ08] definition in two aspects: First, it—in principle—preserves Hou's spectral residual definition [HZ07]. Second, it relies on the combination of the eigenaxes and eigenangles instead of the combination of a single unit pure quaternion and the corresponding phase spectrum (see [GZ10, Eq. 16] and [GMZ08, Eq. 20]), see Fig. 3.5 for an illustration.

(a) Images

(b) PQFT

(c) EPQFT

Fig. 3.5 Example images (**a**) that illustrate the difference between PQFT (**b**) and our EPQFT (**c**) saliency maps. Example images from the Bruce/Toronto (1st and 3rd image, *left* to *right*), Judd/MIT (2th and 5th) and Kootstra (4th) dataset

Quaternion DCT Image Signature Saliency

The signum function for quaternions can be considered as the quaternion's "direction" and is defined as follows:

$$\text{sgn}(x) = \begin{cases} \frac{x_0}{|x|} + \frac{x_1}{|x|}i + \frac{x_2}{|x|}j + \frac{x_3}{|x|}k & \text{for} \quad |x| \neq 0 \\ 0 & \text{for} \quad |x| = 0. \end{cases} \tag{3.34}$$

Given that definition, we can transfer the single-channel definition of the DCT signature and derive the visual saliency S_{QDCT} using the quaternion DCT signature

$$S_{\text{QDCT}} = \mathscr{S}_{\text{QDCT}}(I_Q) = g * \left[T(I_Q) \circ \bar{T}(I_Q) \right] \quad \text{with} \tag{3.35}$$

$$T(I_Q) = \mathscr{D}_Q^{-L}(\text{sgn}(\mathscr{D}_Q^L(I_Q))), \tag{3.36}$$

where again h_{σ_S} is a smoothing Gauss filter with standard deviation σ_S.

3.2.1.4 Weighted Quaternion Components

As proposed by Bian et al. [BZ09], and related to the recent trend to learn feature dimension weights (see, e.g., [ZK11]), we can model the relative importance of the color space components for the visual saliency by introducing a quaternion component weight vector $w = [w_1 \ w_2 \ w_3 \ w_4]^\mathsf{T}$ and adapting Eq. 3.18 appropriately:

$$I_Q = w_4 I_4 + w_1 I_1 i + w_2 I_2 j + w_3 I_3 k . \tag{3.37}$$

In case of equal influence of each color component, i.e. uniform weights, Eq. 3.18 is a scaled version of Eq. 3.37, which is practically equivalent for our application.

3.2.1.5 Multiple Scales

The above spectral saliency definitions only consider a fixed, single scale (see, e.g., [BZ09, GZ10, GMZ08, HHK12]). But, the scale is an important parameter when calculating the visual saliency and an integral part of many saliency models (see, e.g., [FRC10]). For spectral approaches the scale is (implicitly) defined by the resolution of the image I_Q (see, e.g., [JDT11]). Consequently, as proposed by Peters and Itti [PI08b], it is possible to calculate a multiscale saliency map S^M by combining the spectral saliency of the image at different image scales. Let I_Q^m denote the quaternion image at scale $m \in M$, then

$$S^M = \mathscr{S}^M(I_Q) = h_{\sigma_M} * \sum_{m \in M} \phi_r(\mathscr{S}(I_Q^m)), \tag{3.38}$$

where ϕ_r rescales the matrix to the target saliency map resolution r and h_{σ_M} is an additional, optional Gauss filter.

3.2.1.6 Evaluation

To evaluate the considered saliency algorithms, we use the following eye tracking datasets: Bruce/Toronto [BT09], Kootstra [KNd08], and Judd/MIT [JEDT09]. As evaluation measure, we rely on the AUC, because it is the most widely applied and accepted evaluation measure.

Datasets

In the last five years, several eye tracking datasets have been made publicly available to evaluate visual attention models (e.g., [KNd08, BT09, CFK09, JEDT09]; see [WS13]). These easily accessible datasets and the resulting quantitative comparability can be seen as the fuel that has lead to the plethora of novel visual saliency algorithms. Most importantly, the datasets differ in the choice of images (see Fig. 3.6), the number of images, and the number of observers. While the first aspect defines what can be evaluated (i.e., are top-down or specific dominant bottom-up influences present?), a higher number of images and observers leads to more robust evaluation

(a)

(b)

(c)

Fig. 3.6 Example images from the visual saliency evaluation datasets. **a** Bruce/Toronto [BT09]. **b** Judd/MIT [JEDT09]. **c** Kootstra [KNd08]

results, because it reduces the influence of "noise". Here, "free-viewing" refers to a scenario in which the human subjects are not assigned with a task that could lead to a substantial influence of top-down attentional cues such as, for example, to drive a car [BSI13a].

Bruce and Tsotsos's "Toronto" dataset [BT09] is probably the most widely-used dataset to evaluate visual saliency models. It contains 120 color images (681×511 px) depicting indoor and outdoor scenes. Two image categories are dominant within the Toronto dataset: street scenes and object shots, see Fig. 3.6a. The dataset contains eye tracking data of 20 subjects (4 s, free-viewing).

Judd et al.'s "MIT" dataset [JEDT09] contains 1003 images (varying resolutions) selected from Flickr and the LabelMe database, see Fig. 3.6b. Accordingly, the dataset contains huge variations in the depicted scenes. However, there are two very frequent image types: images that depict landscapes and images that show people. The images (variable resolution) were shown to 15 subjects for 3 s with a 1 s gray screen between each two. Eye tracking data was recorded for 15 subjects (3 s, free-viewing).

Kootstra et al.'s dataset [KNd08] contains 100 images (1024×768 px; collected from the McGill calibrated color image database [OK04]). It contains images from five image categories, close-up as well as landscape images, and images with and without a strong photographer bias, see Fig. 3.6c. This substantial data variability makes it particularly hard, because it is difficult to find algorithms that perform well on all of these image types. The images were shown to 31 subjects (free-viewing).

Evaluation Measure

New saliency evaluation measures are proposed regularly and existing measures are sometimes adapted (e.g., [AS13, RDM+13, BSI13b, ZK11, JEDT09, PIIK05, PLN02]). Riche et al. [RDM+13] group visual saliency evaluation measures into three classes. First, value-based metrics such as, for example, normalized scanpath saliency [PIIK05, PLN02], the percentile metric [PI08a], and the percentage of fixations [TOCH06]. Second, several metrics that rely on the area under the receiver operator characteristic curve (e.g., [JEDT09, ZK11, BSI13b]), all of which fall into the group of location-based metrics. Third, there exist distribution-based metrics such as, for example, the correlation coefficient [JOvW+05, RBC06], the Kullback-Leibler divergence [MCB06, TBG05], and the earth mover's distance [JDT12]. Of these evaluation measures, the dominating and most widely applied evaluation measure is the bias-correcting AUC, which—most importantly—has the distinct advantage that it compensates spatial dataset biases (an aspect that we will encounter again in Chap. 4).

The shuffled, bias-correcting area under the receiver operator characteristic curve (AUROC) calculation (see, e.g., [HHK12])—commonly referred to as the AUC evaluation measure—tries to compensate for biases such as, e.g., the center-bias that is commonly found in eye tracking datasets. To this end, we define a positive and a negative set of eye fixations for each image, see Fig. 3.7. The positive sample set contains the fixation points of all subjects on that image. The negative sample set contains the union of all eye fixation points across all other images from the same dataset. To calculate the AUROC, we can threshold each saliency map and the resulting binary map can be seen as being a binary classifier that tries to classify positive and negative samples. Sweeping over all thresholds leads to the ROC curves and defines the area under the ROC curve. When using the AUROC as a measure, the chance level is 0.5 (random classifier), values <0.5 indicate negative correlation, values >0.5 represent positive correlation, and an AUROC of 1 means perfect classification. For eye-fixation prediction the maximally achievable, ideal AUROC is typically substantially lower than 1 (e.g., ∼0.88 on the Bruce/Toronto dataset, ∼0.62 on Kootstra, and ∼0.90 on Judd/MIT). The ideal AUROC is calculated by predicting the fixations of one individual using the fixations of other individuals on the same image. In is necessary to say that the calculation of an ideal AUROC requires a Gaussian filter step. Accordingly, the results in the literature can slightly differ due to different filter parameters and should be seen to serve as guiding values of estimated upper baselines. In some publications, authors normalize the results based on the chance and ideal AUROC values (e.g., [ZK11]). However, the most common practice is to report the original AUROC results and, consequently, we follow this established convention in the following. Thus, when interpreting the results, it is important to consider that the actual value range is practically limited by chance at 0.5 (lower baseline) and the ideal AUROC (upper baseline).

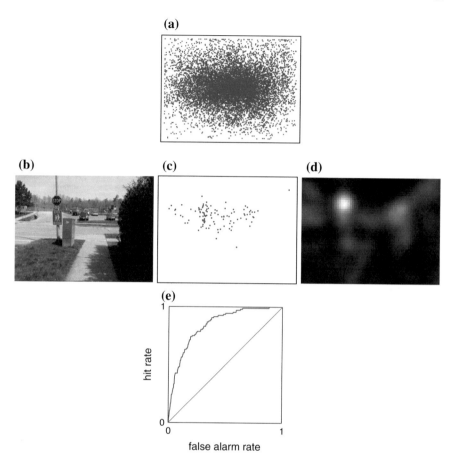

Fig. 3.7 Example image from the Bruce/Toronto dataset to illustrate the components involved when calculating the shuffled, bias-corrected AUC evaluation measure. **a** Negative fixations, **b** image, **c** positive fixations, **d** saliency map, **e** ROC curve

Baseline Algorithms and Results

Table 3.1 shows the results of several baseline algorithms on the three datasets that do not involve face detection, i.e. Kootstra, Judd/MIT, and Bruce/Toronto. The algorithms are: Itti and Koch's model [IKN98] as implemented by the Harel et al. (IK'98) and additionally by Itti's iLab Neuromorphic Vision Toolkit (iNVT'98), Harel et al.'s graph-based visual saliency (GBVS'07; [HKP07]), Bruce and Tsotsos's attention using information maximization (AIM'09; [BT09]), Judd et al.'s linear support vector machine (SVM) approach (JEDA'09; [JEDT09]), Goferman et al.'s context-aware saliency (CAS'12; [GZMT12]), and Lu and Lim's color histogram saliency (CCH'12; [LL12]). Please note that you can find results for further algorithms in, for example,

Table 3.1 AUC Performance of well-known visual saliency algorithms on the three most-commonly used benchmark datasets

	Toronto	Kootstra	Judd
CAS'12	0.6921	0.6033	0.6623
CCH'12	0.6663	0.5838	0.6481
JEDA'09	0.6249	0.5497	0.6655
AIM'09	0.6663	0.5747	0.6379
GBVS'07	0.6607	0.5586	0.5844
IK'98	0.6455	0.5742	0.6365
iNVT'98	0.5442	0.5185	0.5365
Chance	0.5	0.5	0.5
Ideal	~0.88	~0.62	~0.90

Borji et al.'s quantitative saliency evaluation papers (e.g., [BSI13b]). Apart from minor differences, the reported results should be comparable, due to the shared underlying evaluation measure implementation.

Algorithms

As real-valued spectral saliency algorithms, we evaluate: Hou et al.'s spectral residual saliency (SR'07; [HZ07]) and its variant spectral whitening, which is also known as pure Fourier transform and was proposed by Guo et al. (PFT'07; [GMZ08, GZ10]). Furthermore, we evaluate Hou et al.'s DCT signature saliency (DCT'11; [HHK12]).

As quaternion-based algorithms, we evaluate: Guo et al.'s original pure quaternion Fourier transform (PQFT'08; [GMZ08]), which is the quaternion-based counterpart of PFT'07. Our own quaternion-based algorithms, i.e.Eigen pure quaternion Fourier transform (EPQFT) (EPQFT), which is related to PFT'07 and PQFT'08, Eigen spectral residual (ESR), which is related to SR'07, and quaternion discrete cosine transform image signature saliency (QDCT), which is the quaternion-based counterpart of DCT'11. A preceding Δ—imagine a stylized image pyramid—marks algorithms that we evaluated with multiple scales.

We evaluate how well the proposed algorithms perform for all color spaces that have been applied in the literature related to spectral saliency detection: RGB (e.g., [HHK12, HZ07]), ICOPP (e.g., [GZ10, GMZ08]), YUV (e.g., [BZ09]), and CIE Lab (e.g., [HHK12]).

Parameters

We kept the image resolution fixed at 64×48 px in the evaluation, because in preparatory pilot experiments this resolution has constantly shown to provide very good results on all datasets and is the resolution most widely used in the literature (see, e.g., [HHK12]). For multiscale approaches 64 × 48 px is consequently the base resolution. For the Gaussian filtering of the saliency maps, we use the fast recursive filter implementation by Geusebroek et al. [GSvdW03]. We optimized the filter parameters for all algorithms.

Results

First, we can see that the performance depends substantially on the dataset, see Table 3.2, which is not surprising given their different ideal AUCs. We can rank the datasets by the maximum area under the ROC curve that spectral algorithms achieved and obtain the following descending order: Bruce/Toronto, Judd/MIT, and Kootstra. This order can most likely be explained with the different characteristics of the images in each dataset. Two image categories are dominant within the Bruce/Toronto dataset: street scenes and objects. Within these categories, the images have relatively similar characteristics. The Judd/MIT dataset contains many images from two categories: images that depict landscapes and images that show people. The second category is problematic for low-level approaches that do not consider higher-level influences on visual attention such as, e.g., the presence of people and faces in images (we will address this aspect in Sect. 3.2.3). This also is the reason why Judd et al.'s JEDA'09 model performs particularly well on this dataset, see Table 3.1. The Kootstra dataset exhibits the highest data variability. It contains five image categories, close-up as well as landscape images, and images with and without a strong photographer bias. Furthermore, we have to consider that the Kootstra dataset has an extremely low ideal AUC of ~0.62. Accordingly, the dataset contains many images in which an image's recorded gaze patterns vary substantially between persons, which drastically limits the achievable performance.

If we compare the influence of color spaces, then RGB is the color space that leads to the worst performance on all datasets. This is interesting, because it is the only color space in our evaluation that does not try to separate luminance from chrominance information. Interestingly, it appears that the performance difference with respect to the other color spaces (i.e., RGB vs Lab, YUV, or ICOPP) is slightly less for quaternion-based approaches than for real-valued approaches. In other words, quaternion-based approaches seem to be able to achieve better results based on RGB than their real-valued counterparts (see, e.g., the results achieved on the Bruce/Toronto dataset; especially, DCT'11 vs. QDCT and PFT'07 vs. EPQFT). Most interestingly, as we have mentioned in Sect. 3.2.1.2, the quaternion axis transformation can decompose the RGB color space into luminance and chrominance components [ES07]. Accordingly, it is likely that—at least for RGB as a basis— quaternion-based algorithms benefit from their ability to create an intermediate color space that separates luminance from chrominance information.

Table 3.2 AUC performance of the evaluated spectral algorithms

Method	Toronto				Kootstra				Judd			
	Lab	YUV	ICP	RGB	Lab	YUV	ICP	RGB	Lab	YUV	ICP	RGB
Optimal color component weights												
ΔQDCT	0.7201	0.7188	0.7174	0.7091	0.6104	0.6125	0.6110	0.6007	0.6589	0.6751	0.6712	0.6622
QDCT	0.7195	0.7170	0.7158	0.7066	0.6085	0.6119	0.6106	0.5994	0.6528	0.6656	0.6623	0.6552
ΔEPQFT	0.7183	0.7160	0.7144	0.7035	0.6053	0.6082	0.6064	0.5963	0.6527	0.6658	0.6617	0.6559
EPQFT	0.7180	0.7137	0.7122	0.7006	0.6058	0.6073	0.6063	0.5934	0.6483	0.6611	0.6568	0.6493
ΔESR	0.7175	0.7153	0.7133	0.7014	0.6050	0.6077	0.6056	0.5941	0.6508	0.6649	0.6603	0.6534
ESR	0.7162	0.7129	0.7112	0.6990	0.6038	0.6068	0.6044	0.5912	0.6467	0.6601	0.6554	0.6470
ΔPQFT'08	0.7085	0.6969	0.6927	0.6930	0.5943	0.5994	0.5922	0.5868	0.6467	0.6503	0.6429	0.6468
PQFT'08	0.7042	0.6881	0.6826	0.6891	0.5930	0.5970	0.5913	0.5861	0.6404	0.6416	0.6379	0.6398
PQFT'08/Bian [BZ09]	0.7035	0.6880	0.6817	0.6884	0.5928	0.5961	0.5911	0.5861	0.6404	0.6411	0.6375	0.6396
Uniform color component weights												
ΔQDCT	0.7191	0.7107	0.7070	0.7088	0.6050	0.6036	0.6078	0.6002	0.6539	0.6648	0.6618	0.6620
QDCT	0.7180	0.7079	0.7039	0.7056	0.6036	0.6005	0.6079	0.5987	0.6517	0.6572	0.6552	0.6551
ΔEPQFT	0.7148	0.7030	0.7024	0.7026	0.6005	0.5963	0.6045	0.5959	0.6490	0.6530	0.6548	0.6556
EPQFT	0.7141	0.7006	0.6982	0.7006	0.5984	0.5939	0.6023	0.5934	0.6461	0.6496	0.6518	0.6491
ΔESR	0.7142	0.7135	0.7006	0.7013	0.6003	0.5951	0.6028	0.5937	0.6477	0.6504	0.6534	0.6531
ESR	0.7132	0.6998	0.6969	0.6988	0.5975	0.5930	0.6007	0.5909	0.6448	0.6486	0.6502	0.6466
ΔPQFT'08	0.7022	0.6925	0.6868	0.6927	0.5803	0.5826	0.5877	0.5850	0.6431	0.6441	0.6380	0.6465
PQFT'08 [GMZ08]	0.6974	0.6858	0.6796	0.6884	0.5788	0.5808	0.5860	0.5846	0.6368	0.6368	0.6271	0.6396

(continued)

Table 3.2 (continued)

Method	Toronto				Kootstra				Judd			
	Lab	YUV	ICP	RGB	Lab	YUV	ICP	RGB	Lab	YUV	ICP	RGB
Non-quaternion spectral algorithms												
DCT'11 [HHK12]	0.7137	0.7131	0.7014	0.6941	0.6052	0.6089	0.6049	0.5907	0.6465	0.6604	0.6556	0.6461
ΔPFT'07 [PI08b]	0.7177	0.7170	0.7079	0.7014	0.6072	0.6107	0.6084	0.5945	0.6502	0.6601	0.6583	0.6523
PFT'07 [GMZ08]	0.7140	0.7120	0.7025	0.6958	0.6057	0.6079	0.6058	0.5908	0.6445	0.6590	0.6572	0.6446
SR'07 [HZ07]	0.7156	0.7144	0.7051	0.6983	0.6059	0.6090	0.6061	0.5916	0.6462	0.6599	0.6573	0.6461

The performance of non-spectral baseline algorithms is presented in Table 3.1

Within each color space and across all datasets the performance ranking of the algorithms is relatively stable, see Table 3.2. We can observe that without color component weights the performance of the quaternion-based approaches may be lower than the performance of their real-valued counterparts. The extent of this effect depends on the color space as well as on the algorithm. For example, for QDCT this effect does not exist on the RGB and ICOPP color spaces and for Lab only on the Kootstra dataset. However, over all datasets this effect is most apparent for the YUV color space. But, the YUV color space is also the color space that profits most from non-uniform quaternion component weights, see Fig. 3.8, which indicates that the unweighted influence of the luminance component is too high. When weighted appropriately, as mentioned before, we achieve the overall best results using the YUV color space. The influence of quaternion component weights is considerable and depends on the color space, see Table 3.2 and Fig. 3.8. As mentioned it is most important for the YUV color space. However, it is also considerable for Lab and ICOPP. Most importantly, we can observe that the best weights are relatively constant over all datasets.

The importance of multiple scales depends on the dataset. The influence is small for the Bruce/Toronto dataset, which can be explained by the fact that the resolution of 64×48 pixels is nearly optimal for this dataset (see [HHK12]). On the Kootstra dataset the influence is also relatively small, which may be due to the heterogeneous image data. The highest influence of multiple scales can be seen on the Judd/MIT dataset (e.g., compare ΔQDCT with QDCT).

With the exception of Judd et al.'s model on the Judd/MIT dataset—as has been discussed earlier—, our quaternion-based spectral approaches are able to perform better than all evaluated non-spectral baseline algorithms (compare Tables 3.1 and 3.2). We achieve the best single-scale as well as multiscale performance with the QDCT approach. With respect to their overall performance we can rank the algorithms as

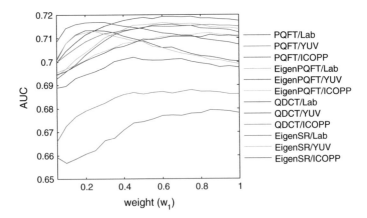

Fig. 3.8 Example of the influence of quaternion color component weights on the AUC performance for QDCT, EPQFT, ESR, and PQFT'08 on the Bruce/Toronto dataset

follows: QDCT, EPQFT, ESR, and PQFT'08. Especially QDCT performs consistently better than its real-valued counterpart DCT'11, see Table 3.2, whereas the situation is not as clear for EPQFT. However, although PFT'07 can achieve slightly better results than EPQFT on the Kootstra dataset, overall EPQFT provides a better performance. Furthermore, we can see that our EPQFT is a substantial improvement over Guo's PQFT'08 and, in contrast to PQFT'08, it is also able to achieve a better performance than the non-spectral baseline algorithms on all three datasets.

In summary, based on a combination of state-of-the-art quaternion-based saliency detection, quaternion component weights, and multiple scales, we are able to improve the state-of-the-art in predicting human eye fixation patterns.

Runtime Considerations

Spectral saliency algorithms inherit the $O(N \log_2 N)$ computational complexity of the discrete fast Fourier transform and in practice also benefit from highly optimized fast Fourier implementations. The (quaternion) FFT- and DCT-based models that we evaluated can be implemented to operate in less than one millisecond (single-scale) on an off the shelf PC. For example, in our implementation of (quaternion) DCT image signatures, we use a hard-coded 64×48 real DCT-II and DCT-III—the latter is used to calculate the inverse—implementation and are able to calculate the bottom-up saliency map in 0.4 ms on an Intel Core i5 with 2.67 GHz (single-threaded; double-precision). This time excludes the time to subsample or resize the image, which depends on the input image resolution, but includes the time for Gauss filtering. This computational efficiency is an important aspect for practical applications and is only a fraction of the computational requirements of most other visual saliency algorithms. For example, assuming a run-time of 1 ms as baseline, the implementations of Judd et al.'s JEDA'09 [JEDT09], Goferman et al.'s CAS'12 [GZMT12], and Bruce and Tsotsos's AIM'09 [BT09] are more than $30,000\times$, $40,000\times$, and $100,000\times$ slower, respectively. Furthermore, our implementation is 20–$50\times$ faster than previously reported for spectral saliency algorithms (see [HHK12, Table II] and [GMZ08, Table 3]) and substantially faster than other run-time optimized visual saliency implementations such as, most importantly, Xu et al.'s multi-GPU implementation of Itti and Koch's saliency model (see [XPKB09, Table II]).

3.2.2 Color Space Decorrelation

As we have seen previously (Sect. 3.2.1), the input color space influences the performance of spectral saliency algorithms. In this context, we noticed that the color spaces that separate lightness from chrominance information (e.g., CIE Lab and YUV) lead to a better performance than RGB. And, we related the relatively good performance of quaternion-based spectral algorithms on the RGB color space to the quaternion axis transformation, which can decompose an RGB image's color information into luminance and chrominance components [ES07]. Interestingly, it is known in research fields such as, e.g., color enhancement that the first principal axis—i.e., the first component after applying the PCA—of an image's or image patch's color information describes the major lightness fluctuations in the scene [GKW87], while the second principal axis describes deviations from the mean color. Furthermore, decorrelation of color information has been successfully applied for several applications such as, e.g., texture analysis and synthesis [LSL00, HB95], color enhancement [GKW87], and color transfer [RP11]. For example, it forms the basis of the well-known decorrelation stretch method to image color enhancement (cf. [All96]).

Interestingly, evidence suggests that specific signals in the human visual system are subject to decorrelation. For example, spatial decorrelation such as lateral inhibition operations is evident in the human vision system. Particularly, this type of spatial decorrelation results in the visual illusion of Mach bands [Rat65], which exaggerates the contrast between edges of slightly differing shades of gray. Buchsbaum and Gottschalk [BG83] and Ruderman et al. [RCC98] found that linear decorrelation of LMS cone responses at a point matches the opponent color coding in the human visual system. Such decorrelation is beneficial for the human visual system, because adjacent spots on the retina will often perceive very similar values, since adjacent image regions tend to be highly correlated in intensity and color. Transmitting this highly-redundant raw sensory information from the eye to the brain would be wasteful and instead the opponent color coding can be seen as performing a decorrelation operation that leads to a less redundant, more efficient image representation. This follows the efficient coding hypothesis of sensory information in the brain [Bar61], according to which the visual system should encode the information presented at the retina with as little redundancy as possible.

We motivated quaternion-based image processing with the wish to being able to process an image's color information holistically, see Sect. 3.2.1 and Fig. 3.3. We did this, because we did not want to process image channels separately and, in consequence, tear apart the color information. Interestingly, we can see color decorrelation as the opposite approach to this holistic idea, because we use decorrelation to make the information that is encoded in the individual color channels as independent or decorrelated as possible. However, this suggests that color decorrelation has the potential to improve the performance of real-valued saliency algorithms that process color image channels separately.

Thus, in the following, we investigate how we use color decorrelation to provide a better feature space for a diverse set of bottom-up, low-level visual saliency algorithms.

3.2.2.1 Decorrelation

Let $I \in \mathbb{R}^{M \times N \times K}$ be the matrix that represents an $M \times N$ image in a color space with K components, i.e. the image has $C = K$ color channels. Reshaping the image matrix and subtracting the image's mean color, we represent the image's mean centered color information in a color matrix $X \in \mathbb{R}^{MN \times K}$.

In general, a matrix W is a decorrelation matrix, if the covariance matrix of the transformed output $Y = XW$ satisfies

$$YY^T = \text{diagonal matrix}. \tag{3.39}$$

In general, there will be many decorrelation matrices W that satisfy Eq. 3.39 and decorrelate [BS97].

The most common approach to decorrelation is the whitening transform, which diagonalizes the empirical sample covariance matrix according to

$$YY^T = C' = WCW^T = I, \tag{3.40}$$

where C is the color covariance matrix

$$C = \frac{1}{MN} \sum_{i=1}^{MN} x_i x_i^T \text{ with } X = \begin{pmatrix} x_1 \\ \cdots \\ x_{MN} \end{pmatrix}. \tag{3.41}$$

Here, C' is the covariance of Y, i.e. of the data after the whitening transform $Y = XW$. As can be seen in Eqs. 3.39 and 3.40, by definition the covariance matrix after a whitening transform equates to the identity matrix, whereas the covariance matrix after an arbitrary decorrelation transform can be any diagonal matrix. Still, there exist multiple solutions for W. The principal component analysis (PCA) computes the projection according to

$$W_{\text{PCA}} = \Sigma^{-1/2} U^T. \tag{3.42}$$

Here, the eigenvectors of the covariance matrix are the columns of Σ and U is the diagonal matrix of eigenvalues. As an alternative, the zero-phase transform (ZCA) [BS97] calculates W according to the symmetrical solution

$$W_{\text{ZCA}} = U \Sigma^{-1/2} U^T. \tag{3.43}$$

The dimensionality preserving color space transform is then given by

$$Y = XW \tag{3.44}$$

and results in the score matrix Y that represents the projection of the image.

Interestingly, the ZCA was introduced by Bell and Sejnowski [BS97] to model local decorrelation in the human visual system. Although the difference in Eqs. 3.42 and 3.43 seems small, the solutions produced by the PCA and ZCA are substantially different (see, e.g., [BS97, Fig. 3]). Interestingly, the ZCA's additional rotation by U, i.e. $W_{\text{ZCA}} = U W_{\text{PCA}}$, causes the whitened data $Y_{\text{ZCA}} = X W_{\text{ZCA}}$ to be as close to the original data as possible.

We reshape the score matrix Y so that it spatially corresponds with the original image and this way obtain our color decorrelated image representation $I_{PCA} \in \mathbb{R}^{M \times N \times K}$. Finally, we normalize each color channel's value range to the unit interval $[0, 1]$. Although not necessary for all saliency algorithms, it is a beneficial step for algorithms that are sensitive to range differences between color components such as, e.g., Achanta's frequency-tuned algorithm [AHES09]. We can then use the decorrelated image channels as a foundation—i.e., in the sense of raw input or feature maps—for a wide range of visual saliency algorithms.

3.2.2.2 Quantitative Evaluation

As in the previous evaluation, we rely on the AUC evalua, and evaluate on the Bruce/Toronto, Kootstra, and Judd/MIT eye tracking datasets (see Sect. 3.2.1.6). Consequently, the baseline results are again shown in Table 3.1. Since color spaces and consequently color decorrelation is such a fundamental aspect that can influence a wide range of algorithms, we take extra precaution to ensure the validity of our claims: First, we evaluate how color space decorrelation influences the performance of eight algorithms. Second, we introduce statistical tests to test the performance of each algorithm on the original color space against the performance based on the image-specific decorrelated color space. Third, although we focus on the AUC evaluation measure in the main body of this book, we do not just rely on the AUC as single evaluation measure in this case and, consequently, we present additional results for the NSS and CC evaluation measures in Appendix C.

At this point, we would like to note that this evaluation's goal is not to assess whether ZCA is a better decorrelation method compared to PCA. Instead, we use two decorrelation methods to indicate that color decorrelation itself is beneficial, independent of a single, specific decorrelation algorithm.

Statistical Tests

We perform statistical significance tests to determine whether or not observed performance differences are significant. Therefore, we record each algorithms prediction for every image and use the evaluation measurements (e.g., AUC) as input data for the statistical tests. We rely on three pairwise, two-sample t-tests to categorize the results: First, we perform a two-tailed test to check whether the compared errors come from distributions with different means (i.e., $\mathcal{H}_=$: "means are equal"). Analogously, second, we perform a left-tailed test to check whether an algorithm's error distribution's mode is greater (i.e., $\mathcal{H}_>$: "mean is greater") and, third, a right-tailed test to check whether an algorithm's error distribution's mode is lower (i.e., $\mathcal{H}_<$: "mean is lower"). All tests are performed at a confidence level of 95 %, i.e., $\alpha = 5 \%$.

Table 3.3 Statistical test
result classes and
visualization color chart

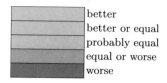

better
better or equal
probably equal
equal or worse
worse

To simplify the presentation and discussion, we group the test results into five classes, see Table 3.3: "Better" means that the hypotheses of equal and worse mean error were rejected. "Better or equal" means that only the hypothesis of a worse mean error could be rejected. "Probably equal" means that no hypothesis could be rejected. "Equal or worse" means that the hypothesis of a better mean error was rejected. "Worse" means that the hypotheses of equal and better mean error were rejected. Here, "better" and "worse" are defined on the desired characteristic or optimum of the target evaluation measure. For example, we would like to maximize the AUROC and accordingly a higher mean is defined as being better.

Algorithms

We adapted the following visual saliency algorithms to evaluate the effect of color space decorrelation: We evaluate Itti and Koch's model (IK'98; [IKN98]) and Harel's graph-based visual saliency (GBVS'07; [HKP07]). For this purpose, we build on Harel's implementation, in which both models share the same groundlying feature maps that can encode color or orientation information. We evaluate pure Fourier transform (PFT'07; [HZ07]) by Hou and Zhang and DCT image signatures (DCT'11; [HHK12]) by Hou et al. Naturally, we also evaluate our own quaternion-based DCT image signatures (QDCT) and EigenPQFT (EPQFT) algorithms, see Sect. 3.2.1. All these algorithms have in common that they are spectral visual saliency algorithms, the first two operate on real-valued images and the latter two process quaternion images. Furthermore, we evaluate the effect on Achanta et al.'s (AC'09; [AHES09]) method, which is based the on each pixel's deviation from the image's mean color. We would like to note that Achanta et al.'s algorithm was developed for salient object detection and not eye fixation prediction. Consequently, we do not expect it to achieve state-of-the-art performance on gaze prediction datasets. Nonetheless, we decided to include Achanta et al.'s algorithm, because we wanted to evaluate a mix of algorithms that rely on different principles for saliency calculation. Additionally, we implemented and adapted Lu and Lim's algorithm (CCH'12; [LL12]) that calculates the visual saliency based on the image's color histogram. Of the above algorithms, IK'98 and GBVS'07 follow the traditional scheme of local center-surround contrast, whereas the spectral approaches (PFT'07, DCT'11, QDCT, and EPQFT), AC'09 and CCH'12 process the image globally.

Parameters

As in our previous evaluation in Sect. 3.2.1.6, we use an image resolution of 64×48 px for spectral saliency approaches. However, in contrast to our previous evaluation, we do not evaluate multiscale approaches, because we have already seen that the integration of multiple scales can further improve the performance. Instead, we focus on the influence of color decorrelation in the following. Therefore, we also use fixed algorithm parameters and do not optimize each algorithm's parameters for each evaluated color space.

Results

We present the achieved results for the Bruce/Toronto, Kootstra, and Judd/MIT dataset in Table 3.4a–c, respectively. To keep our main evaluation compact and readable, we only present the results for RGB, CIE Lab, and ICOPP as base color spaces and base our discussion on the AUC evaluation measure. Results for further color spaces (e.g., Gauss [GvdBSG01] and LMS [SG31]) and evaluation measures (NSS and CC) are presented in Appendix C. Without going into any detail, the results on these additional color spaces and evaluation measures follow the trend that is visible in Table 3.4a–c and thus further substantiate our claim that color space decorrelation is an efficient and robust preprocessing step for many low-level saliency detection algorithms.

As can be seen, the performance of all saliency algorithms improves, if we perform a color space decorrelation. This is independent of the base color space. Even in cases where our statistical tests do not indicate that color space decorrelation improves the results, the mean AUC based on the decorrelated color space is still slightly higher in all cases. Although both evaluated decorrelation methods perform very well, ZCA seems to be slightly better than PCA, because the ZCA leads to the best performance in 44 cases whereas PCA leads to the best performance in 26 cases (the performance is identical for Bruce/Toronto, PFT'07, and Lab). However, this also seems to depend on the saliency algorithm, because DCT'11 and GBVS'07 appear to benefit more from PCA, since PCA leads to the better performance in 7 of 9 cases for DCT'11 and all 9 of 9 cases for GBVS'07.

Interestingly, color space decorrelation leads to better results than quaternion component weighting on the Bruce/Toronto and Kootstra datasets and roughly equal performance on the Judd/MIT dataset. However, although quaternion-based spectral approaches (QDCT and EPQFT) benefit from color space decorrelation, their real-valued counterparts (DCT'11 and PFT'07) seem to provide a slightly better performance in combination with color decorrelation.

In summary, we strongly suggest to perform color space decorrelation for saliency algorithms, because it can significantly increase the performance while it only requires modest computational resources as we will see in the following.

Table 3.4 Color space decorrelation results as quantified by the AUC evaluation measure

(a) Bruce/Toronto dataset

AUC Method	RGB			Lab			ICOPP		
	raw	PCA	ZCA	raw	PCA	ZCA	raw	PCA	ZCA
CCH'12	0.6661	**0.7031**	0.6974	0.6979	0.7061	**0.7072**	0.6881	**0.7032**	0.7019
EPQFT	0.7003	0.7142	**0.7158**	0.7154	0.7180	**0.7212**	0.7112	0.7118	**0.7156**
PFT'07	0.6952	**0.7196**	0.7135	0.7141	**0.7226**	0.7226	0.7128	0.7179	**0.7189**
QDCT	0.7033	**0.7157**	0.7149	0.7158	0.7187	**0.7210**	0.7135	0.7140	**0.7175**
DCT'11	0.6915	**0.7196**	0.7121	0.7126	**0.7208**	0.7207	0.7114	**0.7184**	0.7166
AC'09	0.5406	0.5608	**0.5780**	0.5541	0.5609	**0.5735**	0.5510	0.5543	**0.5702**
GBVS'07	0.6030	**0.6620**	0.6614	0.6371	**0.6665**	0.6655	0.6374	**0.6637**	0.6617
IK'98	0.6410	0.6723	**0.6772**	0.6612	0.6734	**0.6814**	0.6636	0.6721	**0.6756**

(b) Kootstra dataset

AUC Method	RGB			Lab			ICOPP		
	raw	PCA	ZCA	raw	PCA	ZCA	raw	PCA	ZCA
CCH'12	0.5838	0.6030	**0.6045**	0.6018	**0.6043**	0.6037	0.6027	0.6040	**0.6042**
EPQFT	0.5955	0.6050	**0.6140**	0.6021	0.6032	**0.6069**	0.6016	0.6050	**0.6070**
PFT'07	0.5936	**0.6180**	0.6147	0.6087	0.6157	**0.6172**	0.6100	0.6159	**0.6190**
QDCT	0.5974	0.6068	**0.6148**	0.6041	0.6049	**0.6088**	0.6045	0.6069	**0.6092**
DCT'11	0.5891	**0.6148**	0.6143	0.6063	0.6126	**0.6147**	0.6074	0.6134	**0.6173**
AC'09	0.5415	0.5509	**0.5633**	0.5464	0.5487	**0.5544**	0.5463	0.5488	**0.5534**
GBVS'07	0.5584	**0.5897**	0.5879	0.5788	**0.5914**	0.5906	0.5764	**0.5912**	0.5901
IK'98	0.5740	0.5951	**0.5965**	0.5882	0.5936	**0.5950**	0.5881	0.5943	**0.5966**

(c) Judd/MIT dataset

AUC Method	RGB			Lab			ICOPP		
	raw	PCA	ZCA	raw	PCA	ZCA	raw	PCA	ZCA
CCH'12	0.6480	0.6696	**0.6708**	0.6674	**0.6733**	0.6722	0.6595	**0.6705**	0.6702
EPQFT	0.6484	0.6590	**0.6621**	0.6579	0.6581	**0.6609**	0.6547	0.6558	**0.6583**
PFT'07	0.6449	**0.6652**	0.6627	0.6597	**0.6653**	0.6650	0.6590	0.6639	**0.6647**
QDCT	0.6517	0.6608	**0.6625**	0.6599	0.6610	**0.6625**	0.6585	0.6593	**0.6613**
DCT'11	0.6440	**0.6641**	0.6608	0.6581	**0.6645**	0.6638	0.6577	**0.6632**	0.6627
AC'09	0.5306	0.5513	**0.5810**	0.5493	0.5514	**0.5592**	0.5452	0.5492	**0.5585**
GBVS'07	0.5846	**0.6343**	0.6327	0.6207	**0.6367**	0.6362	0.6162	**0.6349**	0.6342
IK'98	0.6367	0.6572	**0.6585**	0.6508	0.6581	**0.6582**	0.6493	0.6556	**0.6564**

Please refer to Table 3.3 for a color legend

Runtime Considerations

We can calculate the PCA in 0.82 ms for a 64×48 color image on an Intel Core i5 with 2.67 GHz (single-threaded; double-precision), in Matlab. Please note that 64×48 px is the default resolution that we use to calculate spectral saliency maps. Here, we use a specialized implementation to calculate the eigenvalues and normalized eigenvectors of hermitian 3×3 matrices based on the Jacobi algorithm. In general, the time to perform the color space decorrelation scales linearly with the number of pixels in the input image. Again, this time excludes the time to subsample or resize the image, which depends on the input image resolution, but includes the time that is required to apply the transformation to the image.

3.2.2.3 Discussion

Given these results, we still have to address what the effects of color decorrelation are and why they can help computational saliency algorithms. For this purpose, we examine the intra and inter color component correlation of color spaces, which is shown for some exemplary color spaces in Table 3.5. Here, the intra color component correlation is the correlation of each color space's individual components (e.g., the correlation of the Lab color space's L and a, L and b, or a and b channels). The inter color component correlation refers to the correlation of the components of different color spaces (e.g., the correlation between RGB's R channel and Lab's a channel).

Does the Decorrelated Color Space Depend on the Input Space?

First of all, the decorrelated color spaces are not independent from their base color spaces. This comes at no surprise, because—for example—an antecedent non-linear transformation such as, e.g., a conversion from RGB to Lab can naturally lead to a different linear decorrelation result, which is illustrated by the low inter component correlation of RGB:PCA and Lab:PCA in Table. 3.5. As a result, we have to neglect the notion of a unique, base color space independent color projection.

Are PCA and ZCA Different?

It becomes apparent in the rightmost column of Fig. 3.9 as well as by the inter color component correlation between RGB:ZCA and RGB or RGB:ZCA and RGB:PCA (see Table 3.5) that ZCA color projections differ substantially from PCA projections, because the ZCA does not seem to separate luminance and chrominance information. This is of interest, because it indicates that not the separation of color and luminance itself is the key to improve the performance, but the properties of decorrelated color information.

Furthermore, we can see that the color components of ZCA projections (e.g., RGB:ZCA or Lab:ZCA) are highly correlated to their base color spaces, see Table 3.5, which stands in contrast to the behavior of PCA projections.

Table 3.5 Mean correlation strength (i.e., absolute correlation value) of color space components calculated over all images in the McGill calibrated color image database [OK04]

	RGB			Lab			ICOPP			LMS			YUV			RGB:PCA			Lab:PCA			RGB:ZCA			Lab:ZCA		
	1st	2nd	3rd	1st	2nd	3rd	1st	2nd	3rd	1st	2nd	3rd	1st	2nd	3rd	1st	2nd	3rd	1st	2nd	3rd	1st	2nd	3rd	1st	2nd	3rd
RGB 1st	1.00	0.88	0.79	0.95	0.32	0.24	0.94	0.39	0.28	0.93	0.86	0.78	0.94	0.28	0.41	0.88	0.05	0.02	0.21	0.07	0.04	0.79	0.45	0.36	0.92	0.26	0.15
RGB 2nd	0.88	1.00	0.89	0.98	0.06	0.09	0.97	0.00	0.16	0.94	0.96	0.88	0.99	0.15	0.03	0.91	0.00	0.03	0.27	0.05	0.00	0.50	0.70	0.47	0.96	0.11	0.03
RGB 3rd	0.79	0.89	1.00	0.88	0.01	0.25	0.93	0.00	0.19	0.86	0.88	0.95	0.90	0.19	0.02	0.89	0.04	0.06	0.36	0.01	0.02	0.40	0.47	0.76	0.91	0.00	0.33
Lab 1st	0.95	0.98	0.88	1.00	0.07	0.14	0.99	0.14	0.19	0.96	0.94	0.86	1.00	0.19	0.16	0.92	0.02	0.07	0.26	0.06	0.02	0.61	0.63	0.45	0.97	0.02	0.06
Lab 2nd	0.32	0.06	0.01	0.07	1.00	0.24	0.10	0.96	0.24	0.10	0.05	0.02	0.07	0.24	0.91	0.08	0.01	0.07	0.10	0.02	0.15	0.73	0.56	0.03	0.05	0.92	0.12
Lab 3rd	0.24	0.09	0.25	0.14	0.24	1.00	0.04	0.42	0.96	0.11	0.05	0.20	0.11	0.99	0.53	0.01	0.11	0.24	0.33	0.13	0.05	0.47	0.28	0.71	0.09	0.11	0.92
ICOPP 1st	0.94	0.97	0.93	0.99	0.10	0.04	1.00	0.15	0.10	0.96	0.95	0.91	0.99	0.10	0.16	0.93	0.00	0.00	0.28	0.05	0.02	0.60	0.57	0.55	0.97	0.06	0.04
ICOPP 2nd	0.39	0.00	0.00	0.14	0.96	0.42	0.15	1.00	0.43	0.16	0.01	0.02	0.13	0.43	0.99	0.12	0.05	0.02	0.15	0.06	0.14	0.80	0.43	0.16	0.11	0.87	0.31
ICOPP 3rd	0.28	0.16	0.04	0.19	0.24	0.96	0.10	0.43	1.00	0.17	0.11	0.15	0.17	0.99	0.54	0.06	0.11	0.24	0.30	0.14	0.05	0.50	0.31	0.68	0.14	0.11	0.89
LMS 1st	0.93	0.94	0.86	0.96	0.10	0.11	0.96	0.16	0.17	1.00	0.98	0.90	0.97	0.16	0.18	0.90	0.00	0.02	0.24	0.05	0.01	0.61	0.58	0.46	0.94	0.05	0.03
LMS 2nd	0.86	0.96	0.88	0.94	0.05	0.05	0.95	0.01	0.11	0.98	1.00	0.93	0.96	0.11	0.03	0.89	0.01	0.02	0.26	0.05	0.01	0.48	0.66	0.50	0.93	0.09	0.02
LMS 3rd	0.78	0.88	0.95	0.86	0.02	0.20	0.91	0.02	0.15	0.90	0.93	1.00	0.88	0.15	0.04	0.86	0.04	0.05	0.33	0.02	0.01	0.38	0.50	0.71	0.88	0.04	0.28
RGB:PCA 1st	0.88	0.91	0.89	0.92	0.08	0.08	0.93	0.12	0.06	0.90	0.89	0.86	0.93	0.06	0.13	1.00	0.00	0.00	0.32	0.01	0.01	0.55	0.53	0.52	0.90	0.05	0.05
RGB:PCA 2nd	0.05	0.00	0.04	0.02	0.01	0.11	0.00	0.05	0.11	0.00	0.01	0.04	0.05	0.11	0.03	0.00	1.00	0.00	0.02	0.21	0.05	0.11	0.00	0.11	0.02	0.07	0.13
RGB:PCA 3rd	0.02	0.03	0.06	0.02	0.07	0.24	0.00	0.02	0.24	0.02	0.02	0.05	0.02	0.24	0.05	0.00	0.00	1.00	0.02	0.04	0.00	0.06	0.27	0.34	0.01	0.15	0.40
Lab:PCA 1st	0.21	0.27	0.36	0.26	0.10	0.10	0.28	0.15	0.30	0.24	0.26	0.33	0.26	0.30	0.19	0.32	0.02	0.02	1.00	0.00	0.00	0.07	0.15	0.29	0.27	0.05	0.20
Lab:PCA 2nd	0.07	0.05	0.01	0.06	0.02	0.02	0.05	0.06	0.14	0.05	0.05	0.02	0.05	0.14	0.08	0.01	0.21	0.04	0.00	1.00	0.00	0.11	0.06	0.07	0.08	0.03	0.12
Lab:PCA 3rd	0.04	0.00	0.02	0.02	0.15	0.15	0.02	0.16	0.05	0.01	0.01	0.01	0.01	0.05	0.01	0.01	0.05	0.00	0.00	0.00	1.00	0.16	0.13	0.03	0.04	0.21	0.00
RGB:ZCA 1st	0.79	0.50	0.40	0.61	0.73	0.47	0.60	0.80	0.50	0.61	0.48	0.38	0.60	0.49	0.82	0.55	0.11	0.06	0.07	0.11	0.16	1.00	0.00	0.00	0.58	0.71	0.38
RGB:ZCA 2nd	0.45	0.70	0.47	0.63	0.56	0.28	0.57	0.43	0.31	0.58	0.66	0.50	0.62	0.32	0.35	0.53	0.21	0.27	0.15	0.06	0.13	0.00	1.00	0.00	0.63	0.69	0.33
RGB:ZCA 3rd	0.36	0.47	0.76	0.45	0.03	0.71	0.55	0.16	0.68	0.46	0.50	0.71	0.48	0.69	0.25	0.52	0.11	0.34	0.29	0.07	0.03	0.00	0.00	1.00	0.51	0.04	0.84
Lab:ZCA 1st	0.92	0.96	0.91	0.97	0.05	0.09	0.97	0.11	0.14	0.94	0.93	0.88	0.98	0.14	0.13	0.90	0.02	0.01	0.27	0.08	0.04	0.58	0.63	0.51	1.00	0.00	0.00
Lab:ZCA 2nd	0.26	0.11	0.00	0.02	0.92	0.11	0.06	0.87	0.11	0.05	0.09	0.04	0.02	0.10	0.82	0.05	0.07	0.15	0.05	0.03	0.21	0.71	0.69	0.04	0.00	1.00	0.00
Lab:ZCA 3rd	0.15	0.03	0.33	0.06	0.12	0.92	0.04	0.31	0.89	0.03	0.02	0.28	0.03	0.90	0.43	0.05	0.13	0.40	0.20	0.12	0.00	0.38	0.33	0.84	0.00	0.00	1.00

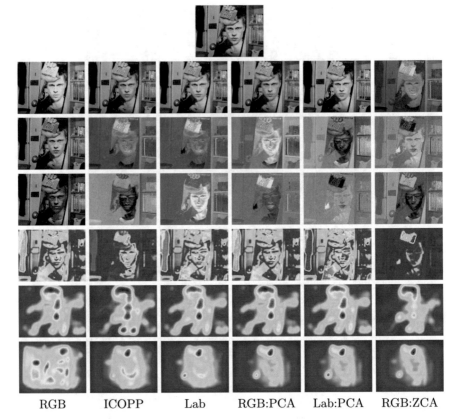

RGB   ICOPP   Lab   RGB:PCA   Lab:PCA   RGB:ZCA

Fig. 3.9 *First row* original image from the Bruce/Toronto dataset; *Rows 2–4* 1st, 2nd, and 3rd color component; *Rows 5–7* saliency maps for AC'09, QDCT, and GBVS'07 (see Sect. 3.2.2.2)

What Is the Effect of Decorrelation?

In fact, there are two aspects of color decorrelation that can influence saliency detection: First, the color information contained in the channels is as decorrelated and thus independent as possible. This naturally supports algorithms that process the color channels independently such as, e.g., DCT'11. Second, algorithms that use color distances (e.g., AC'09) benefit from the aspect that color decorrelation can enhance the contrast of highly correlated images, which is the foundation of the well-known decorrelation stretch color enhancement algorithm (see [GKW87]). In case of the PCA, this is due to the fact that the stretched and thus expanded color point cloud in the decorrelated space is less dense and spread more evenly over a wider volume of the available color space (see, e.g., [GKW87]).

Does Intra Color Component Correlation Influence the Performance
of Saliency Algorithms?

In general, color spaces exhibit different degrees of intra color component correlation, an aspect that is illustrated in Fig. 3.10 and apparent in Table 3.5. Here, our image-specific color decorrelation forms an extreme case—i.e., the intra color component correlation is zero—for which we have demonstrated that it can significantly increase the performance with respect to its base color space. Since it has been noted by several authors that the choice of color spaces directly influences the performance of visual saliency detection algorithms (e.g., [HHK12]), an interesting question is whether or not these performance differences could be related to the color space's degree of intra color component correlation.

To address this question, we calculated the mean performance over all evaluated visual saliency algorithms and the mean intra color component correlation for six well-known color spaces. As we can see in Fig. 3.11, there seems to exist a relation between the average saliency detection performance and the underlying color space's correlation. To quantify this observation, we calculate the correlation between the intra color component correlation and the mean AUC, which is −0.8619, −0.9598, and −0.9067 on the Bruce/Toronto, Kootstra, and Judd/MIT dataset, respectively. Such an overall high negative correlation indicates that a lower visual saliency

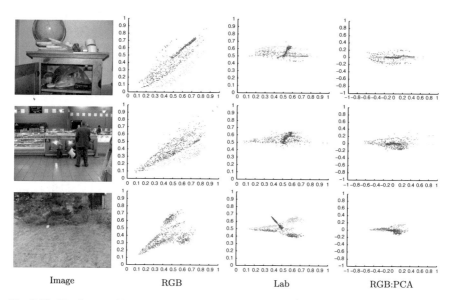

Image   RGB   Lab   RGB:PCA

Fig. 3.10 The degree of intra color component correlation is indicated by the angle of rotation of each point cloud's mean axis. Rotations close to 0° or 90° indicate uncorrelated data and rotations in between indicate degrees of correlation (*red* 1st vs 2nd component; *green* 1st vs 3rd; *blue* 2nd vs 3rd). Visualization method according to Reinhard et al. [RAGS01]. Original images from the Bruce/Toronto dataset

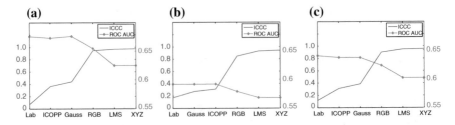

Fig. 3.11 Mean gaze prediction performance of the evaluated visual saliency algorithms (see Sect. 3.2.2.2) and the intra color component correlation (ICCC) for several well-known color spaces. **a** Bruce/Toronto. **b** Kootstra. **c** Judd/MIT

detection performance can be related to a higher intra color component correlation of the underlying color space.

3.2.3   Modeling the Influence of Faces

Up until this point, we have only considered the influence that low-level image features have on visual saliency. However, it has been shown that the gaze of human observers is attracted to faces, even if faces are not relevant for their given task [SS06]. The visual attraction of faces and face-like patterns can already be observed in infants as young as six weeks, which means that infants are attracted by faces before they are able to consciously perceive the category of faces [SS06]. This suggests nothing less than that there exists a bottom-up attentional mechanism for faces [CFK08]. This comes at no surprise, since the perception of the caregivers' faces is an important aspect in early human development, especially for emotion and social processing [KJS+02]. For example, observing the caregiver's face also means to observe the caregiver's eyes and consequently eye gaze, which is essential to start to follow the gaze direction (cf. Sects. 2.1.3 and 4.1). And, following the gaze direction while a caregiver talks about an object in the infant's environment is important, because the relation between gaze and objects is one of the early cues that allow a child to slowly associate spoken words with objects—a key ability to being able to learn a language.

Accordingly, we want to integrate the influence of faces into our computational bottom-up visual saliency model. To this end, we rely face detection and a Gaussian face map to model the presence of faces in an image. For this purpose, without going into any detail, we use Fröba and Ernst's face detection algorithm [FE04] that relies on the modified census transform (MCT) and is known to provide high performance face detections in combination with a very low false positive rate in varying illumination conditions. The output of the face detection algorithm is a set of bounding boxes, each of which reflects the position, size, and orientation of a detected face in the image.

3.2.3.1 Face Detection and the Face Conspicuity Map

In Cerf et al.'s model [CHEK07, CFK09], each detected face is modeled in the face conspicuity map by a circular 2D Gaussian weight function with a standard deviation of $\sigma = \sqrt{(w+h)/4}$, where w and h is the width and height, respectively, of Viola-Jones face detection's bounding box. We extend this model in two ways: First, we allow an in-plane rotation θ of the face bounding boxes provided by our modified census transform (MCT) detectors. Then, we use an elliptical 2D Gaussian weight function g_0, where σ_u and σ_v is the standard deviation in the direction parallel and orthogonal, respectively, to the orientation θ:

$$g_0(u, v, \sigma_u, \sigma_v) = \frac{1}{\sqrt{2\pi\sigma_u}} \exp\left\{-\frac{1}{2}\frac{u^2}{\sigma_u^2}\right\}$$

$$* \frac{1}{\sqrt{2\pi\sigma_v}} \exp\left\{-\frac{1}{2}\frac{v^2}{\sigma_v^2}\right\}, \tag{3.45}$$

where the u-axis corresponds to the direction of θ and the v-axis is orthogonal to θ, i.e.

$$\begin{pmatrix} u \\ v \end{pmatrix} = \begin{pmatrix} \cos\theta & \sin\theta \\ -\sin\theta & \cos\theta \end{pmatrix}\begin{pmatrix} x \\ y \end{pmatrix} = \begin{pmatrix} \hat{\theta}(x) \\ \hat{\theta}(y) \end{pmatrix}. \tag{3.46}$$

Accordingly, we can calculate the face conspicuity map S_F

$$S_F(x, y) = \sum_{1 \leq i \leq N} g_0(\hat{\theta}(x - x_i), \hat{\theta}(y - y_i), \sigma_{u,i}, \sigma_{v,i}, \theta), \tag{3.47}$$

where (x_i, y_i) is the detected center of face i with orientation θ_i and the standard deviations $\sigma_{u,i}$ and $\sigma_{v,i}$. Since, depending on the detector training, the width and height of the bounding box may not be directly equivalent to the optimal standard deviation, we calculate σ_u and σ_v by scaling w and h with the scale factors s_w and s_h that we experimentally determined for our MCT detectors.

3.2.3.2 Integration

Interpreting the calculated visual saliency map S_V and the face detections represented in S_F as two separate low-level modalities, we have to consider several biologically plausible multimodal integration schemes (cf. [OLK07]):

Linear

We can use a linear combination

$$S_+ = w_V S_V + w_F S_F \tag{3.48}$$

as applied by Cerf et al. [CHEK07, CFK09]. However, in contrast to Cerf et al., we analyze the weight space in order to determine weights that provide optimal performance in practical applications. Therefore, we normalize the value range of the saliency map S_V and use a convex combination, i.e. $w_V + w_F = 1$ with $w_V, w_F \in [0, 1]$. From an information theoretic point of view, the linear combination is optimal in the sense that the information gain equals the sum of the unimodal information gains [OLK07].

Sub-linear (Late Combination)

When considering a late combination scheme, no true crossmodal integration occurs. Instead, the candidate fixation points from the two unimodal saliency maps compete against each other. Given saliency maps, we can use the maximum to realize such a late combination scheme, resulting in a sub-linear combination

$$S_{\max} = \max \{S_V, S_F\} . \tag{3.49}$$

Supra-linear (Early Interaction)

Early interaction assumes that there has been crossmodal sensory interaction at an early stage, before the saliency computation and focus of attention selection, which imposes an expansive non-linearity. As an alternative model, this can be realized using a multiplicative integration of the unimodal saliency maps

$$S_\circ = S_V \circ S_F. \tag{3.50}$$

Quaternion Face Channel

From a technical perspective, if the image's color space has less than 4 channels, we can also use the quaternion scalar part to explicitly represent faces and obtain an integrated or holistic quaternion-based saliency map

$$S_Q = \mathscr{S}_{\text{QDCT}}(I_{QF}) \quad \text{with} \tag{3.51}$$
$$I_{QF} = S_F + I_Q = S_F + I_1 i + I_2 j + I_3 k. \tag{3.52}$$

3.2.3.3 Evaluation

Dataset

To evaluate the integration of faces and face detection, we rely on Cerf et al.'s Cerf/FIFA dataset [CFK09]. The dataset consists of eye tracking data (2 s, free-viewing) of 9 subjects for 200 (1024 × 768 px) images of which 157 contain one or more faces, see Fig. 3.12. Additionally, the dataset provides human annotations of the location and size of faces in the images, which can be used to evaluate the influence between perfect, i.e. manual, and automatic face detection.

Procedure

To use the annotated face masks, see Fig. 3.12, as input to our and Cerf's face model, we calculate the principal directions and size of each binary face region. For this purpose, we fit a 2D ellipse that has same normalized second central moments (i.e., spatial variance) as the region. Then, we use the ellipse's major axis length as the face's height and its minor axis length as the face's width; i.e., we assume that a typical face's height is longer than its width. Furthermore, we assume that the ellipse's rotation is identical to the face's orientation.

Algorithms

Since graph-based visual saliency (GBVS) was reported to perform better than Itti and Koch's model [IKN98] when combined with face detections [CHEK07], we compare our system to GBVS'07. As an additional baseline, we include the results reported by Zhao and Koch [ZK11], who used an optimally weighted Itti-Koch model with center bias. We refrain from reporting the evaluation results for all previously

(a)

(b)

Fig. 3.12 Example images from the Cerf/FIFA dataset with their annotated face segments. **a** Images. **b** Masks

evaluated saliency algorithms on the Cerf/FIFA dataset (see Sects. 3.2.1.6, 3.2.2.2, and Appendix C), because we would like to focus the evaluation on the integration of faces and, most importantly, we have already shown the state-of-the-art performance of spectral saliency detection, see Sect. 3.2.2. We report the results for QDCT, EPQFT, PFT'07, and DCT'11. The spectral saliency resolution is set to 64×48 pixels, we rely on the Lab color space with ZCA decorrelation. and the Gaussian filter's standard deviation is 2.5. The standard deviation was set based on the results of a preliminary experiment, in which we independently optimized the spectral saliency filter parameters and the face model parameters.

Results

It can be seen in Table 3.6 that the face map itself has a considerable predictive power, which confirms the observation made by Cerf et al. [CHEK07]. In a few instances, we can even observe that the AUC is higher when using automatic, MCT-based face detection instead of optimal, annotated bounding boxes calculated from the manually annotated face regions. This can be explained by the fact that false positives usually occur on complex image patches that are also likely to attract the attention. Accordingly, false positives do not necessarily have a negative impact on the evaluation results. The linear combination of the bottom-up visual saliency and the face conspicuity map substantially improve the results and we achieve the best results with our scaled elliptical Gauss model. If we look at the results for the two best integration schemes, i.e. linear and late, we can see that our adapted face model is better in all cases but one (GBVS with late combination and face annotations).

If we use the ideal, i.e. human, AUROC to calculate the normalized normalized area under the receiver operator characteristic curve (nAUROC) of our best result with MCT face detections, we obtain an nAUROC of 0.978 which is also higher than the most recently reported 0.962 by Zhao and Koch [ZK11, see Table 1].

The chosen multimodal integration scheme has a considerable influence on the performance, see Fig. 3.13 and Table 3.6. The linear combination achieves the best performance, which is closely followed by the late integration scheme. The integration of the face conspicuity map in the quaternion image does not perform equally well. However, it still substantially outperforms the supra-linear combination, which performs worse than each unimodal map. This could be expected, because the supra-linear combination implies a logical "and".

There is one question that we would like to discuss further: Is the linear combination scheme significantly better than the late combination scheme? To address this question, we resort to our array of statistical tests, see Sect. 3.2.2.2. Unfortunately, the question can not be answered easily and definitely. If we look at the results that we achieve with groundtruth annotations, we see that the results are mixed with beneficial cases for both integration schemes. For example, for QDCT late integration is beneficial in combination with Cerf's face model while linear integration is beneficial for our face model. In both example cases the statistical tests leave not much room for interpretation. The p-values for our three t-tests (i.e., higher, equal, and lower

Table 3.6 AUC performance of the evaluated algorithms on the Cerf/FIFA dataset [CFK09]

Face detection	Annotated		MCT	
Face model	Cerf	Our	Cerf	Our
Linear combination				
EPQFT	0.7601	0.7697	0.7641	0.7685
PFT'07	0.7577	0.7677	0.7610	0.7666
QDCT	0.7611	**0.7706**	0.7634	**0.7686**
DCT'11	0.7593	0.7682	0.7615	0.7673
GBVS'07	0.7223	0.7306	0.7120	0.7284
Late combination				
EPQFT	0.7632	0.7689	0.7529	**0.7617**
PFT'07	0.7594	**0.7667**	0.7487	0.7606
QDCT	0.7660	**0.7667**	0.7445	0.7613
DCT'11	0.7624	0.7655	0.7434	0.7600
GBVS'07	0.7238	0.7089	0.6632	0.7047
Early interaction				
EPQFT	0.6581	0.6582	0.6355	0.6660
PFT'07	0.6586	0.6586	0.6365	0.6667
QDCT	0.6588	0.6589	0.6373	0.6666
DCT'11	**0.6593**	0.6588	0.6372	**0.6670**
GBVS'07	0.6537	0.6575	0.6366	0.6632
Quaternion face channel				
EPQFT	0.7111	0.7115	0.7118	0.7110
QDCT	**0.7140**	0.7138	0.7145	**0.7142**
Face-only				
Faces	0.6566	**0.6594**	0.6367	**0.6648**
Saliency-only				
EPQFT	0.7223			
QDCT	0.7205			
DCT'11	0.7204			
PFT'07	0.7199			
GBVS'07	0.6350			
Further baseline				
Zhao and Koch*, 2011	0.7561			

The ideal, i.e. human, AUC is 0.786. *We used the ideal AUC to calculate our AUC for Zhao and Koch's reported result [ZK11]

mean) are close to (0, 0, 1) in the first case and (1, 0, 0) in the second case. Here, a potential cause might be the sometimes distorted groundtruth segmentation masks, e.g. see the two rightmost images in Fig. 3.12. However, linear integration is better in all cases for MCT face detection. Given that the performance differences appear

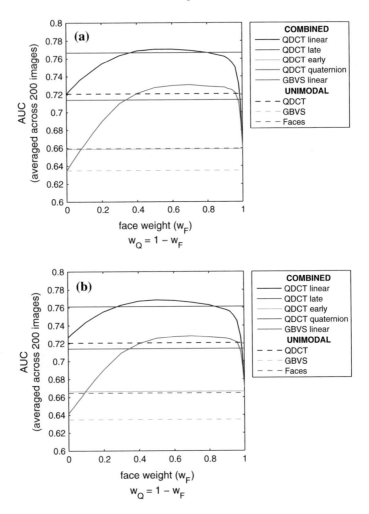

Fig. 3.13 Illustration of the average AUC in dependency of the chosen face integration method on the Cerf/FIFA dataset [CFK09]. **a** Our face model based on face annotations. **b** Our face model based on MCT face detections

quite substantial for MCT detections, it comes at no surprise that the statistical tests indicate that linear integration is in fact "better" for all these cases, i.e. if we rely on MCT face detection. In combination with the fact that the performance is better for a relatively large value range of w_F, see Fig. 3.13, we can only suggest to use the linear integration for practical applications with automatic face detections.

3.3 Auditory Attention

Two fundamental concepts are involved in human bottom-up auditory attention that form the basis for our computational attention model (cf. Sect. 2.1.2): First, auditory attention relies on audio data that is subject to a frequency analysis that is realized by the basilar membrane. Second, the brain's auditory attention system relies—among other aspects—on so-called novelty detection neurons that encode deviations from the pattern of preceding stimuli. We can model the first concept with common time-frequency analysis methods (Sect. 3.3.1.1). Here, it is interesting that by doing so, we can rely on computations that can in later stages be reused by other auditory tasks such as, e.g., speech recognition or sound source localization. To model the second concept, we can assign a "surprise" neuron to each frequency, following Itti and Baldi's theory of Bayesian surprise [IB06]. Such neurons probabilistically learn and adapt to changes of each frequency's distribution over time. To detect novel, odd, or changed signal components, we can then measure how far a newly observed sample deviates from our learned pattern (Sect. 3.3.1.2), which in principle is similar to the brain's signal mismatch detection mechanism.

3.3.1 Auditory Novelty Detection

In sensory neuroscience, it has been suggested that only unexpected information is transmitted from one stage to the next stage of neural processing [RB99]. According to this theory, the sensory cortex has evolved neural mechanisms to adapt to, predict, and suppress expected statistical regularities [OF96, MMKL99, DSMS02] to focus on events that are unpredictable and appear as being novel, odd, or "surprising". It is intuitively clear that "surprising" signals and events can only occur in undeterministic environments. This means that surprise arises from the presence of uncertainty that can be caused by, for example, intrinsic stochasticity or missing information. Interestingly, it has been shown in probability and decision theory that the Bayesian theory of probability provides the only consistent and optimal theoretical framework to model and reason about uncertainty [Jay03, Cox64]. Accordingly, Itti and Baldi [IB06] suggested a Bayesian approach to model neural responses to surprising signals, see Fig. 3.14.

In the Bayesian probability framework, probabilities correspond to subjective degrees of beliefs (see, e.g., [Gil00]) in models that are updated according to Bayes' rule as new data is observed. According to Bayesian surprise, the background information of an observer is represented in the prior probability distribution $\{P(M)\}_{M \in \mathcal{M}}$ over the models M in a model space $\mathcal{M}$. Given the prior distribution of beliefs, a new data observation D is used to update the prior distribution $\{P(M)\}_{M \in \mathcal{M}}$ into the posterior distribution $\{P(M|D)\}_{M \in \mathcal{M}}$ via Bayes' rule

$$\forall M \in \mathcal{M}: \quad P(M|D) = \frac{P(D|M)}{P(D)} P(M). \tag{3.53}$$

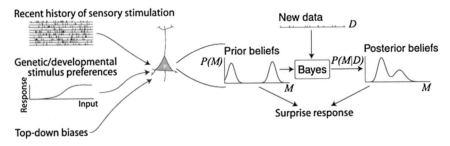

Fig. 3.14 Bayesian surprise as a probabilistic model for novelty detection on the basis of simple neurons. Image from [IB09] reprinted with permission from Elsevier

In this framework, the new data observation D carries no surprise, if it leaves the observer beliefs unaffected, i.e. the posterior is identical to the prior. D is surprising, if the posterior distribution after observing D significantly differs from its prior distribution. To formalize this, Itti and Baldi propose to use the Kullback-Leibler divergence (KLD) to measure the distance D_{KL} between the prior and posterior distribution

$$S(D, \mathcal{M}) = D_{KL}(P(M|D), P(M)) = \int_{\mathcal{M}} P(M|D) \log \frac{P(M|D)}{P(M)} dM. \quad (3.54)$$

The distance $S(D, \mathcal{M})$ between the prior and posterior now quantifies how surprising observation M is.

Since the posterior distribution can be updated immediately after observing new data, it is clear that surprise is an attention model that is particularly suited to detect

Fig. 3.15 An approximately ten second exemplary audio sequence in which a person speaks and places a solid object on a table at the end of the sequence. The measured audio signal (**a**), the resulting Gaussian auditory surprise (**b**), and the spectrogram (**c**)

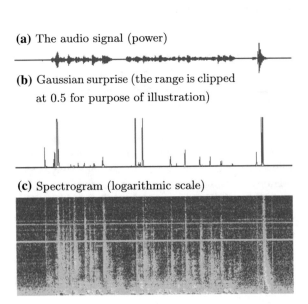

(a) The audio signal (power)

(b) Gaussian surprise (the range is clipped at 0.5 for purpose of illustration)

(c) Spectrogram (logarithmic scale)

surprising events online and without delay in sensory streams such as, most importantly, audio and video streams.

3.3.1.1 Time-Frequency Analysis and Bayesian Framework

We can use the short-time Fourier transform (STFT), short-time cosine transform
(STCT), or the modified discrete cosine transform (MDCT) to calculate the spectro-
gram $G(t, \omega) = |F(t, \omega)|^2$ of the windowed audio signal $a(t)$ (see Fig. 3.15c), where
t and ω denote the discrete time and frequency, respectively. Accordingly, at each
time step t, the newly observed frequency data $G(t, \omega)$ is used to update the prior
probability distribution

$$\forall \omega \in \Omega : \quad P_{\text{prior}}^{\omega} = P(G(\cdot, \omega)|G(t-1, \omega), \ldots, G(t-N, \omega)) \quad (3.55)$$

of each frequency and obtain the posterior distribution

$$\forall \omega \in \Omega : \quad P_{\text{post}}^{\omega} = P(G(\cdot, \omega)|G(t, \omega), G(t-1, \omega), \ldots, G(t-N, \omega)), \quad (3.56)$$

where $N \in \{1, \ldots, \infty\}$ allows additional control of the time behavior by limiting the
history to $N \neq \infty$ elements, if wanted. The history allows us to limit the influence
of samples over time and consequently "forget" data, which is essential for the time
behavior of the Gaussian surprise model.

3.3.1.2 Auditory Surprise

Gaussian Model

Using the Gaussian distributions as model, we can calculate the auditory surprise
$S_A(t, \omega)$ for each frequency

$$S_A(t, \omega) = D_{\text{KL}}(P_{\text{post}}^{\omega}||P_{\text{prior}}^{\omega}) = \int P_{\text{post}}^{\omega} \log \frac{P_{\text{post}}^{\omega}}{P_{\text{prior}}^{\omega}} dg \quad (3.57)$$

$$= \frac{1}{2}[\log \frac{|\Sigma_{\text{prior}}^{\omega}|}{|\Sigma_{\text{post}}^{\omega}|} + \text{Tr}\left[\Sigma_{\text{prior}}^{\omega^{-1}} \Sigma_{\text{post}}^{\omega}\right] - I_{\text{D}}$$

$$+ (\mu_{\text{post}}^{\omega} - \mu_{\text{prior}}^{\omega})^T \Sigma_{\text{prior}}^{\omega^{-1}}(\mu_{\text{post}}^{\omega} - \mu_{\text{prior}}^{\omega})] \quad , \quad (3.58)$$

where μ and Σ is the mean and variance, respectively, of the data in the considered
time window. D_{KL} is the KLD and Eq. 3.58 results from the closed form of D_{KL} for
Gaussian distributions [HO07].

Gamma Model

The Gaussian model is extremely run-time efficient and in general performs well according to our experience. But, it has one main disadvantage: All elements inside the history window have equal weight. Instead of equal weights, it would be desirable that the weight and thus the influence of each observation slowly decreases over time to realize a "smooth" forgetting mechanism. Similar to the approach by Itti and Baldi for detecting surprising events in video streams [IB05], we can use the Gamma distribution as an alternative to the Gaussian distribution

$$P(x) = \gamma(x; \alpha, \beta) = \frac{\beta^\alpha x^{\alpha-1} e^{-\beta x}}{\Gamma(\alpha)} \tag{3.59}$$

with $x \geq 0$, $\alpha, \beta > 0$, and Gamma function Γ to calculate the surprise.

Given a new observation $G(t, \omega)$ and prior density $P^\omega_{\text{prior}} = \gamma(\cdot; \alpha, \beta)$, we calculate the posterior $P^\omega_{\text{post}} = \gamma(\cdot; \alpha', \beta')$ using Bayes' rule

$$\alpha' = \alpha + G(t, \omega) \tag{3.60}$$

$$\beta' = \beta + 1. \tag{3.61}$$

However, using this update rule would lead to an unbounded growth of the values over time. To avoid this behavior and reduce the relative importance of older observations, we integrate a decay factor $0 < \zeta < 1$

$$\alpha' = \zeta\alpha + G(t, \omega) \tag{3.62}$$

$$\beta' = \zeta\beta + 1. \tag{3.63}$$

This formulation preserves the prior's mean $\mu = \frac{\alpha}{\beta} = \frac{\zeta\alpha}{\zeta\beta}$ but increases its variance, which however represents a relaxation of belief in the prior's precision after observing $G(t, \omega)$.

Now, we can calculate the surprise as follows

$$S_A(t, \omega) = D_{\text{KL}}(P^\omega_{\text{post}} \| P^\omega_{\text{prior}}) = \int P^\omega_{\text{post}} \log \frac{P^\omega_{\text{post}}}{P^\omega_{\text{prior}}} dg \tag{3.64}$$

$$= \alpha' \log \frac{\beta}{\beta'} + \log \frac{\Gamma(\alpha')}{\Gamma(\alpha)} \tag{3.65}$$

$$+ \beta' \frac{\alpha}{\beta} + (\alpha - \alpha')\psi(\alpha), \tag{3.66}$$

where ψ is the Digamma function. Unfortunately, the Gamma and Digamma functions Γ and ψ, respectively, do not have a closed form. But, there exist sufficiently accurate approximations (see, e.g., [Ber76]), which however make the calculation slightly more complex than in the case of the Gaussian model.

3.3.1.3   Across Frequency Combination

Finally, we calculate the auditory saliency $S_A(t)$ as the mean over all frequencies (see Fig. 3.15b)

$$S_A(t) = \frac{1}{|\Omega|} \sum_{\omega \in \Omega} S_A(t, \omega). \tag{3.67}$$

We do not use an alternatively possible joint (e.g., Dirichlet) model for the surprise calculation due to its computational complexity. Such a joint model would require the calculation of a general covariance matrix with every update. Given the typically large number of analyzed frequencies (i.e., >10000), the associated computational complexity makes real-time processing impractical if not impossible.

3.3.2   Evaluation

In contrast to, for example, recording eye fixations as a measure of visual saliency (see Sect. 2.1.1), we can not simply observe and record humans to provide a measure of auditory saliency. Consequently, we follow a pragmatic, application-oriented evaluation approach that enables us to use existing acoustic event detection and classification datasets.

3.3.2.1   Evaluation Measure

Salient acoustic event detection has to suppress "uninteresting" audio data while highlighting potentially relevant and thus salient acoustic events. However, in contrast to classical acoustic event detection and classification, this consideration leads to a different evaluation methodology in which: First, a high recall is necessary, because we have to detect all prominent events so that they can be analyzed by later processing stages. Second, a high precision is of secondary interest, because we can tolerate false positives as long as we still filter the signal in such a way that we achieve a net benefit when taking into account subsequent processing stages. We can realize this evaluation idea by using the well-established F_β score

$$F_\beta = (1 + \beta^2) \cdot \frac{\text{precision} \cdot \text{recall}}{(\beta^2 \cdot \text{precision}) + \text{recall}} \tag{3.68}$$

$$F_\beta = \frac{(1 + \beta^2) \cdot \text{true pos.}}{(1 + \beta^2) \cdot \text{true pos.} + \beta^2 \cdot \text{false neg.} + \text{false pos.}} \tag{3.69}$$

as evaluation measure, where β "measures the effectiveness of retrieval with respect to a user who attaches β times as much importance to recall as precision" [vR79]. It is noteworthy that F_β is also the most commonly used evaluation measure for salient object detection in image processing applications (see Sect. 4.2.3.B).

3.3.2.2   Evaluation Data

We use the CLEAR2007 acoustic event detection dataset for evaluation [CLE] (cf. [TMZ+07]). The dataset contains recordings of meetings in a smart room and its collection was supported by the European integrated project Computers in the Human Interaction Loop (CHIL) and the US National Institute of Standards and Technology (NIST). For each recording a human analyst marked and classified all occurring acoustic events that were remarkable enough to "pop-out" from the acoustic scene's background noise. In total, 14 different classes of acoustic events were classified and flagged, including sudden "laughter", "door knocks", "phone ringing" and "key jingling". Here, it is interesting to note that not all events could be identified by the human analyst, in which case they were labeled with "unknown".

3.3.2.3   Evaluation Parameters

For the time-frequency analysis, we set the window size to contain 1 second of audio data, which has a resolution of 22 kHz, and use 50 % overlap. We also experimented with applying various window functions (e.g., Blackman, Gauss), but the resulting performance difference is mostly negligible, if the window functions' parameters are well defined. We evaluated the performance for the modified discrete cosine transform (MDCT), short-time cosine transform (STCT), and short-time Fourier transform (STFT) to determine whether or not the Gamma distribution is beneficial for all of these transformations. We do this, because one aim is to produce as little run-time overhead as possible, which requires us to ideally rely on the transformation that is used for the subsequent processing steps such as, e.g., sound source localization, event recognition, and/or speech recognition. We optimized the history size and forgetting parameter for the Gaussian and Gamma model, respectively, and report the results for the best choice.

3.3.2.4   Results

As can be seen in Table 3.7, quantified using the F_1, F_2, and F_4 score, auditory surprise is able to efficiently detect arbitrary salient acoustic events. Although in general an F_1 score of roughly 0.77 is far from perfect for precise event detection, we can see from the substantially higher F_2 and F_4 scores that we can efficiently detect most (salient) acoustic events, if we tolerate a certain amount of false positives. This nicely fulfills the target requirements for our application domain and comes at a low computational complexity, since Gaussian surprise allows us to process one minute of audio data in roughly 1.5 s. This makes it possible to process the incoming

Table 3.7 Performance of the evaluated auditory surprise algorithms on CLEAR 2007 acoustic event detection data

Algorithm	F_1	F_2	F_4
STFT + Gamma	0.7668	0.8924	0.9665
STCT + Gamma	0.7658	0.8916	0.9655
MDCT + Gamma	0.7644	0.8894	0.9647
STFT + Gaussian	0.7604	0.8832	0.9531
STCT + Gaussian	0.7612	0.8813	0.9529
MDCT + Gaussian	0.7613	0.8805	0.9538

The F_2 and F_4 scores are our main evaluation measure, because for our application a high recall is much more important than a high precision (we provide the F_1 score mainly to serve as a reference). We can see that surprise is able to reliably detect arbitrary, interesting acoustic events

audio data stream in real-time, detect salient events online, and signal occurring salient events to subsequent stages with a minimum delay. Furthermore, since we calculate the surprise value for all frequencies that we subsequently combine, we can also determine which frequencies trigger the detection. An information that can be passed to subsequent stages to focus the processing on these frequencies. We can see in Table 3.7 that the Gamma distribution leads to a better performance compared to the Gauss distribution, independently of the preceding time-frequency transformation. This, however, comes at the cost of greater computational complexity.

3.4 Saliency-Based Audio-Visual Exploration

In the previous sections, we have investigated saliency models to determine what attracts the auditory or visual attention (Sects. 3.2 and 3.3, respectively). However, to realize an attention system that sequentially focuses on salient regions in the scene— similar to human saccades—, we need to define a crossmodal representation, extract auditory and visually salient regions, sequentially shift the focus of attention, and keep track of attended objects to implement inhibition of return (cf. Sects. 2.1.1 and 2.1.3).

Attention forms a selective gating mechanisms that decides what will be processed by later stages. This process is often describes as a "spotlight" that enhances the processing in the attended [TG80, Pos80], i.e. "illuminated", region. In a similar metaphor, attention can act like a "zoom lense" [ESJ86, SW87], because the size of the attended region can be adjusted depending on the task. However, most models do not consider the shape and extent of the attended object, which is essential to determine the area that has to be attended. And, experimental evidence suggests that attention can be tied to objects, object parts, and/or groups of objects [Dun84, EDR94, RLS98]. But, how can we attend to objects before we recognize them [WK06]?

One model that addresses this question has been introduced by Rensink [Ren00a, Ren00b]. Rensink describes "proto-objects" as volatile units of visual information that can be bound into a coherent and stable object when accessed by focused attention [WK06]. A related concept that we have already addressed earlier (see Sect. 2.1.1) are Kahneman and Treisman's "object files" [KT00, KTG92]. The main difference between proto-objects and object files is the role of location in space. In Kahneman and Treisman's object file model, the spatial location is just another property of an object, i.e. it is just another entry in the object's file, see Fig. 2.1. In contrast, in Rensink's proto-object model and coherence theory (see [SY06]), the spatial location serves an index that binds together various low-level features into proto-objects across space and time [Ren00a, WK06].

3.4.1   Gaussian Proto-Object Model

We rely on a probabilistic model to represent salient auditory, visual, and audio-visual proto-objects. In our model, every proto-object $o \in \mathcal{O}$ is represented by a parametric Gaussian weight function

$$f_o^G(x) = \frac{s_o}{\sqrt{(2\pi)^3 \det(\Sigma_o)}} \exp\left(-\frac{1}{2}(x - \mu_o)^T \Sigma_o^{-1}(x - \mu_o)\right) \qquad (3.70)$$

with $x \in \mathbb{R}^3$. Here, the 3D mode $\mu_o \in \mathbb{R}^3$ represents the likely spatial center of the proto-object, the variance Σ_o reflects the spatial extent that means—more generally—the spatial area that likely contains the actual object, and s_o is the proto-object's saliency. In accordance to our Gaussian model, we can represent every proto-object o as a 3-tuple h_o

$$h_o = (s_o, \mu_o, \Sigma_o) \in \mathcal{H}. \qquad (3.71)$$

3.4.2   Auditory Proto-Objects

In Sect. 3.3, we have discussed how we determine how salient a sound signal is at time t. However, in order to form a proto-object, we have to determine the coarse spatial area that likely contains the salient signal's sound source.

3.4.2.1   Localization

We rely on the well-known steered response power (SRP) with phase transform (PHAT) sound source localization [MMS+09, DSB01]. The SRP-PHAT algorithm uses the inter-microphone time difference of arrival (TDOA) of sound signals, which is caused by the different distances the sound has to travel to reach each microphone, to estimate the location of the sound source. To this end, the following inter-microphone signal correlation function is used to determine TDOAs τ of prominent signals at time t

$$R_{ij}(t, \tau) = \int_{-\infty}^{\infty} \psi_{ij}^{\text{PHAT}}(t, \omega) F_i'(t, \omega) F_j'(t, \omega)^* e^{j\omega\tau} d\omega, \qquad (3.72)$$

where F_i' and F_j' are the STFT transformed signals of the audio signal at microphone i and j, respectively. The PHAT specific weighting function

$$\psi_{ij}^{\text{PHAT}}(t, \omega) = |F_i'(t, \omega) F_j'(t, \omega)^*|^{-1} \qquad (3.73)$$

can be regarded as a whitening filter and is supposed to decrease the influence of noise and reverberations. Subsequently, we can use the estimated TDOAs to calculate the corresponding spatial positions in the environment.

3.4.2.2 Parametrization

Since the sound source localization is a process that exhibits a considerable amount of noise, we perform spatio-temporal clustering to remove outliers and improve the accuracy of the localization. Accordingly, we can use the mean of each cluster as the proto-object's location estimate μ_o and calculate the corresponding co-variance matrix Σ_o. Consequently, each detected acoustically salient proto-object o is described by its saliency s_o, the estimated location μ_o, and the co-variance matrix Σ_o that encodes the spatial uncertainty.

3.4.3 *Visual Proto-Objects*

In Sect. 3.2, we have described how we calculate the visual saliency of an image. However, to represent this information in our proto-object model, we have to determine 2D proto-object regions in the saliency map. Then, we can use depth information that can be provided by, for example, stereo vision to form a 3D proto-object representation.

3.4.3.1 Proto-Object Regions

We analyze the saliency map's isophote curvature (see [LHR05]) to estimate the proto-object regions, see Fig. 3.16. Here, isophotes are closed curves of constant saliency within the saliency map. Assuming a roughly (semi-)circular structure of

(b) The attentional shifts for the 10 most-salient proto-objects based on inhibition of return

(a) Exemplary scene image

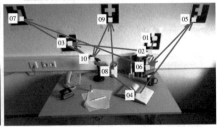

(d) Saliency map with fitted Gaussians proto-object regions

(c) Saliency accumulator map

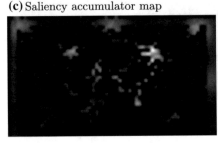

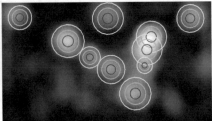

Fig. 3.16 An exemplary scene image (**a**), the saliency map with fitted Gaussian proto-object regions (**d**), the resulting accumulator (**c**), and the first 10 salient proto-object regions that are selected by the location-based inhibition of return (**b**). The estimated Gaussian weight descriptors are depicted as overlay on the saliency map (illustrated as *circles* with center μ_o and radii $r \in \{\sigma_o, 2\sigma_o, 3\sigma_o\}$ in $\{red, green, yellow\}$, respectively). Please note that the value range of the saliency map and the accumulator is attenuated for purpose of illustration

salient peaks, we can then determine the center of each salient peak as well as its corresponding pixels that we define as the pixels whose gradients point toward the peak center. This way, we are able to efficiently extract salient regions, even in the presence of highly varying spatial extent and value range, partially overlapping peaks and noise. To this end, we analyze the local isophote curvature κ of the visual saliency map S_V

$$\kappa = -\frac{S_{cc}}{S_g} = -\frac{S_y^2 S_{xx} - 2S_x S_y S_{xy} + S_x^2 S_{yy}}{(S_x^2 + S_y^2)^{3/2}} \quad, \tag{3.74}$$

where S_{cc} is the second derivative in the direction perpendicular to the gradient and S_g is the derivative in gradient direction.[3] Accordingly, S_x, S_y and S_{xx}, S_{xy}, S_{yy} are the first and second derivatives in x and y direction, respectively. Exploiting that the local curvature is reciprocal to the (hypothetical) radius r of the circle that generated the saliency isoline of each pixel, i.e.

---

[3]Please note that all operations in Eqs. 3.74 and 3.76 operate element-wise. We chose this simplified notation for its compactness and readability.

$$r(x, y) = \frac{1}{\kappa(x, y)},\qquad(3.75)$$

we can estimate the location of each peak's center. Therefore, we calculate the displacement vectors (D_x, D_y) with

$$D_x = \frac{S_x \left(S_x^2 + S_y^2\right)}{S_{cc}} \quad \text{and} \quad D_y = \frac{S_y \left(S_x^2 + S_y^2\right)}{S_{cc}}\qquad(3.76)$$

and the resulting hypothetical peak centers (C_x, C_y) with

$$C_x = P_x - D_x \quad \text{and} \quad C_y = P_y - D_y,\qquad(3.77)$$

where the matrices P_x and P_y represent the pixel abscissae and ordinates, i.e. the pixel (x, y) coordinates, respectively.

Thus, we can calculate a saliency accumulator map A_S in which each pixel votes for its corresponding center. The most salient regions, i.e. corresponding to the extents of the proto-objects in the image (see, e.g., [HZ07]), in the saliency map can then be determined by selecting the pixels of the accumulator cells with the highest voting score, see Fig. 3.16c. By choosing different weighting schemes for the voting, we are able to implement divers methods for assessing the saliency of each region. In the following, we use the saliency as weight and normalize each accumulator cell $A_S(m, n)$ by division by the number of pixels that voted for the pixel, i.e.

$$A_S(m, n) = \frac{\sum_x \sum_y 1_m(C_x(x, y)) 1_n(C_y(x, y)) S_V(x, y)}{\sum_x \sum_y 1_m(C_x(x, y)) 1_n(C_y(x, y))},\qquad(3.78)$$

where $1_x(y)$ is the indicator function with $1_x(y) = 1$ iff $x = y$ and $1_x(y) = 0$ otherwise. However, due to noise and quantization effects, we additionally select pixels that voted for accumulator cells within a certain radius r, i.e.

$$A_S'(m, n) = \frac{\sum_x \sum_y 1_{m,n}^r(C_x(x, y), C_y(x, y)) S_V(x, y)}{\sum_x \sum_y 1_{m,n}^r(C_x(x, y), C_y(x, y))} \quad \text{with}\qquad(3.79)$$

$$1_{m,n}^r(x, y) = \begin{cases} 1 & \text{if } \sqrt{(m - x)^2 + (m - y)^2} \leq r \\ 0 & \text{otherwise.} \end{cases}\qquad(3.80)$$

Unfortunately, the initially selected pixels of our proto-object regions are contaminated with outliers caused by noise. Thus, we perform convex peeling (cf. [HA04]), a type 1, unsupervised clustering-based outlier detector to remove scattered outliers and eliminate regions whose percentage of detected outliers is too high.

To extract all salient proto-object regions that attract the focus of attention, we apply a location-based inhibition of return mechanism on the saliency map (see, e.g., [RLB+08, IKN98]; [SRF10, SF10a]). To this end, we use the accumulator to select the most salient proto-object region and inhibit all pixels within the estimated outline by setting their saliency to zero. This process is repeated until no further prominent salient peaks are present in the map.

3.4.3.2 Parametrization

For each extracted 2D salient proto-object region $o \in \mathcal{O}_{2D}(S_V(t))$ within the visual saliency map S_V at time t, we derive a parametric description by fitting a Gaussian weight function f_o. We assume that the Gaussian weight function encodes two distinct aspects of information: the saliency s_o as well as the (uncertain) spatial location and extent of the object μ_o and Σ_o, respectively. Consequently, we decompose the Gaussian weight function:

$$f_o^G(x) = \frac{s_o}{\sqrt{(2\pi)^D \det(\Sigma_o)}} \exp\left(-\frac{1}{2}(x - \mu_o)^T \Sigma_o^{-1}(x - \mu_o)\right) \qquad (3.81)$$

with $D = 2$. Exploiting a stereo setup or other RGB-D sensors, we can estimate the depth and project the 2D model into 3D. This way, we obtain a 3D model for each visually salient proto-object region that follows the representation of the detected auditory salient events, see Sect. 3.4.2.2. However, we have to make assumptions about the shape, because the spatial extent of the object in direction of the optical axis can not be observed. Thus, we simplify the model and assume a spherical model in 3D and, accordingly, a circular outline in 2D, i.e. $\Sigma_o = I_D \sigma_o$ with the unit matrix I_D.

3.4.4 Audio-Visual Fusion and Inhibition

3.4.4.1 Saliency Fusion

After the detection and parametrization of salient auditory and visual signals, we have a set of auditory $\mathcal{H}_A$ and visual $\mathcal{H}_V$ proto-objects represented in a Gaussian notation at each point in time t

$$\{h_1, \ldots, h_N\} = \mathcal{H}^t = \mathcal{H}_A^t \cup \mathcal{H}_V^t, \qquad (3.82)$$

where each proto-object $o_i \in \mathcal{O}$ is represented by a 3-tuple h_i consisting of its saliency s_{o_i}, spatial mean μ_{o_i}, and spatial variance Σ_{o_i}, see Eq. 3.71. To reduce the influence of noise as well as to enable multimodal saliency fusion, we perform a cross-modal spatio-temporal mean shift clustering [CM02] of the auditory and visual Gaussian representatives. Accordingly, we obtain a set of audio-visual clusters $C^t \in \mathcal{P}(\mathcal{H}^t)$, each of which can be interpreted as a (saliency-weighted) Gaussian mixture model. Therefore, we interpret each cluster $c \in C^t$ as a saliency-weighted Gaussian mixture model, that consists of auditory and/or visual proto-objects. This allows us to split each cluster c again into an auditory ($c_A = c \cap \mathcal{H}_A^t$) and/or visual ($c_V = c \cap \mathcal{H}_V^t$) sub-cluster and estimate the saliency for each modality separately. Subsequently, we consider a linear combination to integrate the audio-visual saliency

$$s_o = \frac{1}{2}\left(\sum_{o_k \in c_A} w^A_{o_k} f^G_{o_k}(\mu_o) + \sum_{o_l \in c_V} w^V_{o_l} f^G_{o_l}(\mu_o) \right), \tag{3.83}$$

using the modality specific weights $w^A_{o_j}$ and $w^V_{o_j}$ (analogous to Eq. 3.84). Consequently, we use the spatial mean of every proto-object within the cluster to estimate the position

$$\mu_o = \mathbb{E}[c] = \sum_{o_j \in c} \mu_{o_j} w_{o_j} \quad \text{with } w_{o_j} = \frac{s_{o_j}}{\sum_{o_i \in c} s_{o_i}}. \tag{3.84}$$

Finally, we determine the spatial variance of the cluster Σ_o by iteratively fusing the variance of the proto-objects

$$V_j = V_{j-1} - V_{j-1}\left(V_{j-1} + \Sigma_{o_j}\right)^{-1} V_{j-1}, \quad \forall_{j=2,\dots,H} \tag{3.85}$$

with $V_1 = \Sigma_{o_1}$, $\Sigma_o = V_H$, and $H = |c|$. Accordingly, we are able to build a new audio-visual proto-object $h_o = (s_o, \mu_o, \Sigma_o)$ with integrated saliency s_o, spatial mean μ_o as well as spatial variance Σ_o.

We use a linear combination for crossmodal integration, because it has been shown to be a good model for human overt attention and is optimal according to information theoretic criteria [OLK07], see Sect. 2.1.3. However, other combination schemes (see, e.g., [OLK07]) can be realized easily given the model and algorithmic framework.

3.4.4.2 Object-Based Inhibition of Return

An important additional feature of our spatio-temporal fusion in combination with the employed object-based world model, see [KBS+10], is the object-centric representation of salient regions. This allows us to use the euclidean distance metric to relate the current proto-objects with previous proto-object detections at previous time steps as well as already analyzed objects that are stored in a world model. This way, we can decide whether to create and attend a new proto-object or update the information of an already existing entity.

To iteratively attend and analyze the objects present in the scene, we use the detected salient proto-objects to realize an object-based inhibition of return mechanism. Therefore, at each decision cycle, the most salient proto-object cluster that is not related with an already attended and analyzed proto-object gains the overt focus of attention.

3.4.4.3 Knowledge-Driven Proto-Object Analysis

After the sensors have been aligned with respect to the proto-object in the current overt focus of attention, the foveal cameras (see Fig. 3.17) are used to inspect the

(a) The ARMAR-III robot head   **(b)** PTU stereo setup

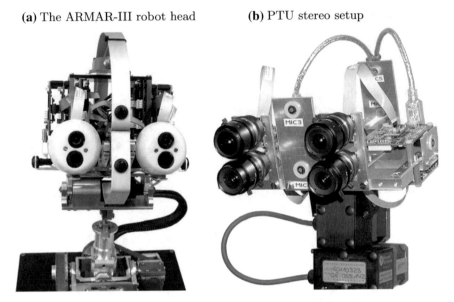

Fig. 3.17 The ARMAR-III humanoid robot head (**a**) and our pan-tilt-unit (PTU) stereo setup (**b**) provide 7 and 2 degrees of freedom, respectively. Both setups perceive their environment with 6 omnidirectional microphones (1 left, 1 right, 2 front, 2 rear) and 2 stereo camera pairs (coarse and fine view, respectively)

object. Therefore, we extend the multimodal knowledge-driven scene analysis and object-based world modeling system as presented by Machmer et al. [MSKK10] and Kühn et al. [KBS+10], to comply with our iterative, saliency-driven focus of attention and exploration mechanism. Most importantly, we replaced the detection and instantiation phase by regarding proto-objects as primitive candidates for world model entities. The attended proto-object region is instantiated as entity and subsequently hierarchically specialized and refined in a knowledge-driven model (see [MSKK10, KBS+10]). The analysis of each proto-object is finished, if no further refinement is possible, which marks the end of the decision cycle and initiates the next shift of attention. Within this framework, every entity is tracked which is an important feature of object-based inhibition of return.

3.4.5 *Evaluation*

3.4.5.1 **Hardware and Software Setup**

The sensor setup that was used for the evaluation of the presented system is shown in Fig. 3.17. The wide angle and foveal cameras have a focal length of 6 and 3.5 mm, respectively. The stereo baseline separation between each camera pair is 90 mm.

The camera sensors provide a resolution of $640 \times 480\,\text{px}$ at a frame rate of 30 Hz. In the evaluation only the front and side omnidirectional microphones are used (see Fig. 3.17). The distance between the side microphones is approximately 190 mm and the vertical distance between the front microphones is approximately 55 mm. The pan-tilt unit provides an angular resolution of $0.013°$ and is mounted on a tripod in such a way that the cameras are roughly on eye height of an averagely tall human to reflect a humanoid view of the scene.

The audio data is processed at a sampling rate of 48 kHz. A Blackman window with a size of 512 samples and 50 % overlap is used to calculate the STFT for the Gaussian auditory surprise, which uses a history size of $N = 128$. In the following, the STFT F' of the sound source localization uses a lower temporal-resolution than the STFT F of the salient event detection. This is due to the fact that we require real-time performance and, on the one hand, want to detect short-timed salient events while, on the other hand, require sufficiently large temporal windows for robust correlations. Therefore, the window length of the localization is a multiple of the salient event detections' window length. Accordingly, we aggregate the auditory saliency of all detection windows that are located within the localization window. We use the maximum as aggregation function, because we want to react on short-timed salient events, instead of suppressing them.

3.4.5.2   Evaluation Procedure and Measure

First of all, to demonstrate that overt attention is beneficial and justifies the required resources, we assess the impact of active sensor alignment on the perception quality (Sect. 3.4.5.3). While the improvement of the image data quality of objects in the focused foveal view compared to the coarse view is easily understandable (see Fig. 3.18), the impact on the acoustic perception depends on several factors, most importantly the sensor setup. Consequently, as reference we evaluate the acoustic localization error with respect to the pan-tilt orientation of our sensor setup relative to sound sources, e.g. household devices and speaking persons. For this purpose, the sound sources were placed at fixed locations and the localization was performed with pan-tilt orientations of $\{-80°, \ldots, 80°\} \times \{-30°, \ldots, 0°\}$ in $10°$ steps (see Fig. 3.19). We only consider the angular error, because in our experience the camera-object distance error is too dependent on the algorithm parameters, implementation, and sampling rate.

Fig. 3.18 Exemplary image of an object in the coarse and fine view, respectively

Fig. 3.19 Mean sound source localization error (in °) depending on the pan-tilt-orientation of the sensor setup

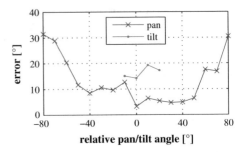

We perform a couple of experiments to evaluate the behavior of the proposed system, because a quantitative, comparative method to evaluate the performance of an overt attention system does not exist (see [SS07, BKMG10]). In order to obtain a reliable impression of the performance of our system, we repeated every experiment multiple times with varying environmental conditions such as, e.g., lighting, number of objects, distracting clutter, and timing of events. Inspired by the evaluation procedures by Ruesch et al. [RLB+08] and Begum et al. [BKMG10], we investigate and discuss the performance of saliency-driven visual and multimodal scene exploration.

3.4.5.3 Results and Discussion

Audio-Visual Perception

As can be seen in the error curve depicted in Fig. 3.19, the angular localization error is minimal if the head faces the target object directly. This can be explained by the hardware setup in which the microphones are nearly arranged on a meridional plane. Interestingly, the curve shows a non-monotonic error progression, which is mainly caused by the hardware that interferes with the acoustic characteristic and perception, e.g. the cameras heavily influence the frontal microphones (see Fig. 3.17). Additionally, in Fig. 3.18 we show an example of the coarse and fine, i.e. foveal, view of a focused object to illustrate the improved visual perception, i.e. increased level of detail.

Visual Exploration I—FoA Shift

In style of the experimental evaluation by Ruesch et al. [RLB+08, Sect. V–B], we mounted three salient calibration markers on the walls of an office environment and removed other distracting stimuli (see Fig. 3.20). In this experiment, we benefit from an object-specific lifetime that can be assigned to analyzed objects in our world model. Each object-specific lifetime is continuously evaluated and updated by, e.g., taking the visibility into account. Thus, if an object has expired and is perceived as salient, it can regain the focus of attention. Driven by the implemented inhibition of

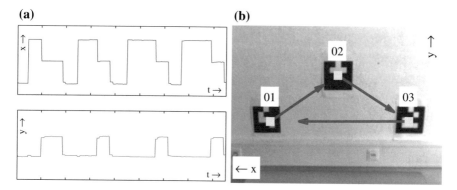

Fig. 3.20 A short temporal section of the attended x-y position (**a**) in the cyclic focus of attention shift experiment (see [RLB+08]). The positions correspond to the calibration marker locations (**b**) that lie on the same x-z plane. **a** Temporal behavior. **b** Marker arrangement

return mechanism, the three salient marks are explored by shifting the overt attention from one mark to the next most salient mark that is not inhibited. As expected, the achieved behavior corresponds to the cyclic behavior as described by Ruesch et al. [RLB+08]. Each attended mark is focused by controlling the pan-tilt-servos. The resulting trajectory is illustrated in Fig. 3.20.

Visual Exploration II—Object-Based IoR

Once an object has been analyzed, it is tracked and inhibited—as long as the object has not been marked for re-focusing by higher-level processes—from gaining the overt focus of attention. To test the object-based inhibition of return mechanism, we perform experiments with moving objects in the scene. For this purpose, we place movable objects in the scene, start the exploration, and move objects after they have been analyzed. As expected, smoothly moving objects do not attract the focus, although they are moved to locations that have not been salient before. Naturally, this behavior even remains when motion is integrated as an additional saliency cue. Interestingly, objects that abruptly change their expected motion pattern attract the focus of attention again, because the tracking of the entity in the object-based world modeling system fails due to the unexpected movement (see Sect. 3.4.4.3). Although this could be seen as a technical deficit, this behavior is desired for an attention-based system and can be biologically motivated (cf. [HKM+09]).

Multimodal Exploration I—FoA Shift

Following the experimental procedure of Ruesch et al. [RLB+08, Sect. V–C], we examine the behavior in scenes with acoustic stimuli. Therefore, we extend the scenario of the previous experiment (Sect. 3.4.5.3) and add a single visible sound source, e.g. a blender or a talking person. Our system explores the environment

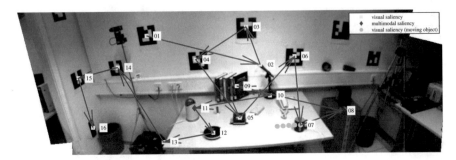

Fig. 3.21 An example of multimodal scene exploration: The focus of attention is shifted according to the numbers in the stitched image of the scene (only the first 15 shifts are shown). The *yellow squares mark* objects that attracted the focus solely due to their visual saliency whereas the *blue squares* (at 08 and 11) mark audio-visually caused shifts of attention. Furthermore, the *green dotted lines* (at 07) roughly indicate the trajectory of the moved object

based on visual saliency until the acoustic stimulus begins and the sound source directly gains the focus of attention.

Multimodal Exploration II—Scene

Finally, we unite the previously isolated experiments and assess the performance on more complex scenes with several objects, object motion, and auditory stimuli (please see Fig. 3.21 for an exemplary scene). The system is capable of handling these situations according to our expectations. Most importantly, objects that are auditory and visually salient tend to attract the saliency even if they are not the most salient point in each modality. Furthermore, salient sound sources outside the visual field of view compete with visually salient stimuli and both are able to attract the overt focus of attention due to the normalized value ranges (see Sect. 3.4.4.1).

3.5 Multiobjective Exploration Path

Iteratively attending the most salient region that has not been attended yet is the classical approach to saliency-based overt and covert attention, see Sect. 3.4. However, in many situations, it is advisable to integrate further target criteria when planning where to look next. For example, it might be interesting to maximize the coverage of previously unseen areas with each attentional shift [MFL+07], integrate top-down target information for visual search ([OMS08]; cf. Chap. 4), or implement a task-dependent spatial bias to specific regions of the environment ([DBZ07]; cf. Chap. 4). In our opinion, it is also beneficial to minimize ego-motion. This has several benefits such as, among others: First, it can reduce the time to focus the next and/or all selected objects. Second, it can save the energy that is required to move joints.

Third, it can reduce wear-and-tear of mechanical parts due to an overall reduction of servo movement. It also has another beneficial side-effect, because it often leads to less erratic and—according to our subjective impression—more human-like head motion patterns compared to saliency-only exploration strategies.

Given the detected salient proto-objects, we can mathematically address the problem of where to look next as an optimization problem, i.e. to determine the order of proto-objects that minimizes a given target function. By adapting the target function toward different criteria, we easily can implement a diverse set of exploration strategies. In the following, we present our balanced exploration approach that realizes a tradeoff between rapid saliency maximization and ego-motion minimization. However, we hope that you will agree with us that given our problem formulation it is easily possible to integrate further target criteria.

3.5.1 Exploration Path

We define an exploration path $EP \in S(\mathcal{O})$, i.e. the order in which to attend the proto-objects, as a permutation of the proto-objects $\{o_1, o_2, \ldots, o_N\} = \mathcal{O}$ that are scheduled to be attended. Here, $S(\mathcal{O})$ is the permutation group of $\mathcal{O}$ with $|\mathcal{O}| = N!$. For example, the exploration path $EP_{example} = (o_1, o_3, o_2, o_4)$ would first attend object o_1, then o_3 followed by o_2, and finally o_4. In the following, we denote s_{o_i} as the saliency of object o_i and q_{o_i} represents the robot's joint angle configuration needed to focus object o_i. Accordingly, the target function that determines the optimal exploration path has the form $f_{target} : S(\mathcal{O}) \to \mathbb{R}$ and in the following we define f_{target} and try to solve for

$$EP_{opt} = \arg \min_{EP \in S(\mathcal{O})} f_{target}(EP). \tag{3.86}$$

3.5.2 Exploration Strategies

3.5.2.1 Saliency-Based Exploration Path

Analog to saliency-only bottom-up exploration as presented in Sect. 3.4, we can sort all perceived proto-objects by their saliency s_{o_i} in descending order and attend the proto-objects in the resulting order $EP_{saliency}$, i.e.

$$EP_{saliency} = (o_{i_1}, o_{i_2}, \ldots, o_{i_N}) \text{ with } s_{o_{i_1}} \geq, \cdots, \geq s_{o_{i_N}}. \tag{3.87}$$

3.5.2.2 Distance-Based Exploration Path

Alternatively, we can ignore the saliency and try to minimize the accumulated joint angle distances that are necessary to attend all selected proto-objects

$$\text{EP}_{\text{distance}} = \arg\min_{EP \in S(\mathcal{O})} \left\{ \sum_{k=1}^{N} \left\| q_{o_{i_k}} - q_{o_{i_{k-1}}} \right\| \right\}, \qquad (3.88)$$

where $q_{o_{i_k}}$ represents the joint angles needed to focus the kth object and $q_{o_{i_{k-1}}}$ is the joint angle configuration for the preceding object. Here, $q_{o_{i_0}}$ is defined as being the initial joint angle configuration at which we start the exploration. We use the norm of the joint angle differences $d_{m,n} = \|q_m - q_n\|$ as a measure for the amount of necessary ego-motion between two joint configurations. Unfortunately, to determine the minimal accumulated distance to attend all proto-objects is an NP-complete problem, because it equates to the traveling salesman problem [CLR90, Weg05].[4] Consequently, we limit the computation to K local neighbors of the currently focused object that were not already attended. In our implementation, we use $K = 10$, which seems to provide good results at acceptable computational costs. This strategy leads to paths that minimize the required amount of ego-motion, but it does not take the saliency into account.

3.5.2.3 Balanced Exploration Path

Considering the exploration path planning as a multiobjective optimization problem [Ehr05], we can combine the saliency-based and distance-based approach. To this end, we define a single aggregate objective function

$$\text{EP}_{\text{balance}} = \arg\min_{EP \in S(\mathcal{O})} \left\{ \sum_{k=1}^{N} f_d(\|q_{o_{i_k}} - q_{o_{i_{k-1}}} \|) \cdot f_s(s_{o_{i_k}}) \right\}, \qquad (3.89)$$

where $s_{o_{i_k}}$ is the saliency value of the proto-object o_{i_k}, f_d is a distance transformation function, and f_s is a saliency transformation function. We define f_d as identity function and $f_s(s; \alpha) = s^{-\alpha}$, i.e.

$$\text{EP}_{\text{balance}}(\alpha) = \arg\min_{EP \in S(\mathcal{O})} \left\{ \sum_{k=1}^{N} \|q_{o_{i_k}} - q_{o_{i_{k-1}}} \| \cdot s_{o_{i_k}}^{-\alpha} \right\}. \qquad (3.90)$$

This aggregate optimization function implements the tradeoff between attending far away proto-objects with a high saliency and nearby proto-objects with a lower

---

[4]Please note that the traveling salesman problem (TSP)'s additional requirement to return to the starting city does not change the computational complexity.

saliency, where the choice of α weights the target objective's priorities. This optimization problem equates to an asymmetric TSP. It is asymmetric, because the aggregate function's distance term depends on the object's saliency, see Eq. 3.90, which leads to a different distance between two joint configurations depending on the end configuration. Accordingly, we search for an approximate solution and limit the search for the next best object to K local neighbors of the currently attended object.

3.5.3 Evaluation

Although it seems impossible to quantitatively evaluate the system behavior, see Sect. 3.4.5.2, we try to approach a quantitative evaluation of the exploration strategies in two steps: First, we record the whole environment in a scan sweep and calculate the locations of all salient proto-objects. This is similar to a person that takes a quick, initial glance around the room to get a first impression of an environment. Second, given a starting configuration, we can use the pre-calculated salient proto-object locations to plan the robot's eye movement. This way, the first step enables us to analyze different methods to determine the salient regions and the second step makes it possible to analyze specific properties of the generated active behavior.

3.5.3.1 Data

We recorded a dataset that consists of 60 videos (30 s each) to evaluate our exploration strategies. The videos were recorded using two hardware platforms in different environments, see Table 3.8, Figs. 3.17 and 3.22. We re-enacted sequences in three scenarios: office scenes, breakfast scenes, and neutral scenes. Here, neutral scenes were recorded in the same environment, but with a reduced amount of salient objects.

3.5.3.2 Evaluation Measures

Since a comparable evaluation has not been performed before, we had to develop novel evaluation measures that allow us to quantitatively compare the presented exploration strategies. We propose two evaluation measures that model different, competing goals.

We use the cumulated joint angle distance (CJAD) as measure of robot ego-motion

$$\text{CJAD(EP)} = \frac{1}{N} \sum_{j=1}^{N} \left\| q_{\text{EP}_j} - q_{\text{EP}_{j-1}} \right\|, \tag{3.91}$$

where EP_j is the index of the jth attended object of exploration path EP, and q_{o_i} represents the joint angle configuration that focuses object o_i, see Sect. 3.5.2.2. Since

Table 3.8 Composition of the exploration path evaluation data

Scene	Number of recordings		
	PTU/sensors	ARMAR-III	total
Breakfast	15	15	30
Office	10	10	20
Neutral	5	5	10
Total	30	30	60

we want to reduce the amount of necessary head motion, we want to minimize the CJAD.

To investigate the influence of saliency on the exploration order, we use the cumulated saliency (CS) of already attended objects

$$\text{CS}(i; \text{EP}) = \sum_{j=1}^{i} s_{\text{EP}_j}, \ i \in \{1, 2, \dots, N\}. \tag{3.92}$$

We want to observe a steep growing curve, because this would mean that objects with higher saliency are attended first. Since the number of attended salient objects may vary depending on the saliency distribution in the scene, we denote the percentage of already attended objects as p, which makes it possible to integrate over the curves of different scenes. This way, we can calculate the area under the CS curve—we refer to it as integrated cumulated saliency (ICS)—as a compact evaluation measure, i.e.

$$\text{ICS}(\text{EP}) = \int \text{CS}(p; \text{EP}) \, dp. \tag{3.93}$$

Furthermore, we introduce normalized cumulated joint angle distance (NCJAD) and normalized cumulated saliency (NCS) as normalized versions of cumulated joint angle distance CJAD and cumulated saliency CS, respectively:

$$\text{NCJAD}(\text{EP}) = \frac{\text{CJAD}(\text{EP}) - \text{CJAD}(\text{EP}_{\text{distance}})}{\text{CJAD}(\text{EP}_{\text{saliency}}) - \text{CJAD}(\text{EP}_{\text{distance}})} \tag{3.94}$$

$$\text{NCS}(\text{EP}) = \frac{\text{ICS}(\text{EP}_{\text{saliency}}) - \text{ICS}(\text{EP})}{\text{ICS}(\text{EP}_{\text{saliency}}) - \text{ICS}(\text{EP}_{\text{distance}})}. \tag{3.95}$$

The advantage of NCJAD and NCS is that they consider the spatial distribution of objects in the scene as well as their saliency distribution. This normalization terms

$$\text{CJAD}(\text{EP}_{\text{saliency}}) - \text{CJAD}(\text{EP}_{\text{distance}}) \quad \text{and} \tag{3.96}$$

$$\text{ICS}(\text{EP}_{\text{saliency}}) - \text{ICS}(\text{EP}_{\text{distance}}) \tag{3.97}$$

are the result of two considerations: First, the saliency-based exploration necessarily leads to the fastest growth of CS and thus highest ICS, but it is likely to have a high

(a) Room 1, PTU

(b) Room 2, ARMAR-III

Fig. 3.22 Sample image stitches of the recordings with the stereo camera pan-tilt-unit (PTU) head (**a**) and with the ARMAR-III head (**b**)

CJAD. Second, the distance-based strategy leads to the smallest CJAD, but is likely to exhibit a slow growth of CS.

To serve as a lower boundary for CS, we calculate $EP_{saliency*}$ which is the opposite strategy to $EP_{saliency}$ that selects the least salient unattended object at each shift. Analogously, we calculate $EP_{distance*}$ which greedily selects the object with the highest distance at each step and is an approximate (greedy) strategy opposite to $EP_{distance}$.

3.5.3.3 Results and Discussion

Exploration Path I—Saliency-Based

First, we examine the saliency-based exploration approach (see Fig. 3.23, red; see Sect. 3.5.2.1) that is most widely found in related work and formed the basis for our qualitative experiments in Sect. 3.4. This strategy leads to a high amount of head movement (high CJAD), in fact the highest of all strategies, but it also leads to the highest growth of the cumulated saliency (high ICS; see Fig. 3.24). This leads to a slower exploration of all objects in the scene, but a fast analysis of the most salient objects.

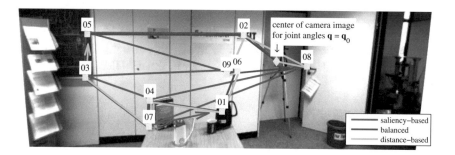

Fig. 3.23 An example to illustrate the different focus of attention selection strategies. The attention shifts for each path are illustrated in the stitched image (only the nine most salient locations are shown and *yellow squares mark* the positions of the objects)

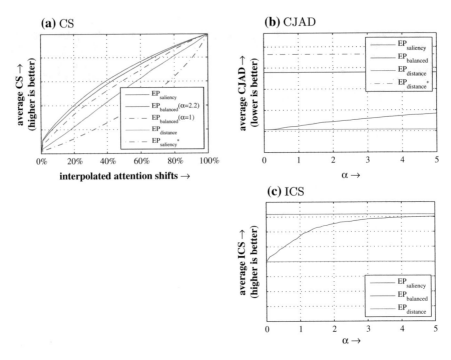

Fig. 3.24 The average cumulated saliency (**a**), the average cumulated joint angle distances (**b**), and the average area under the cumulated saliency curve (**c**) over all recordings in the database

Exploration Path II—Distance-Based

Second, we investigate the exploration strategy that minimizes the angular distances (see Fig. 3.23, green; see Sect. 3.5.2.2). Compared to the other strategies, the resulting exploration pathes do not take into account saliency of proto-objects, which leads to the slowest cumulated saliency growth (low ICS; see Fig. 3.24). But, as can be seen in Fig. 3.24, the necessary angular distances and thus the time required for the full scene exploration is minimized (low CJAD). We would like to note that the computational limitation of using only K local neighbors for the TSP optimization (see Sect. 3.5.2.2) leads to a 25 % longer distance in general [JM97].

Exploration Path III—Balanced

Finally, we consider the balanced strategy that implements a tradeoff between a small cumulated joint angle distance and steep growth of cumulated saliency (see Fig. 3.23, blue; see Sect. 3.5.2.3). We can adjust the priority of these two competing goals by changing the operating parameter α. Interestingly, even a relatively high α can already significantly reduce the CJAD while providing a high ICS, see Fig. 3.24.

When α is set to 2.2, we achieve an average CJAD of 0.2972. For comparison the distance-based and saliency-based strategy achieve a CJAD of 0.2157 (72.6 %) and 0.7576 (254.9 %), respectively. At the same time, we achieve an average ICS of 156.5. Here, the distance-based and saliency-based strategy achieve an average ICS of 130.0 (83.1 %) and 161.8 (103.4 %), respectively. Thus, we provide an exploration strategy that effectively balances between favoring highly salient objects and efficient head movements.

3.6 Summary and Future Directions

We presented how we integrate saliency-driven, iterative scene exploration into a hierarchical, knowledge-driven audio-visual scene analysis approach. In principle, this follows the idea by Treisman et al., see Sect. 2.1.1, and—to our best knowledge—has not been done to this extent by any other research group. To realize this system, we had to overcome several obstacles in different areas that we will recapitulate in the following.

When we started to work on audio-visual saliency-based exploration in 2010, we faced the situation that many methods that have been developed around the field of computational attention were not suited for use in real robotic systems. This was caused by the fact that most methods—including the saliency models themselves—were ill-suited for our use case, computationally too complex, or simply not state-of-the-art. Furthermore, only one comparable audio-visual robotic attention system existed [RLB+08], which relied on comparatively simple models and methods.

Although computational auditory attention seems to attract an increasing interest (e.g., [NSK14, RMDB+13, SPG12]), still only few auditory saliency models exist (most importantly, ours and [KPLL05, Kal09]). And, the models that existed were computationally demanding and not suited for online processing. But, online processing was a necessary requirement for being able to immediately detect and react on interesting acoustic events (e.g., a shattering glass or a person starting to speak). Having this goal in mind, we developed auditory surprise, which uses a Bayesian model to efficiently detect acoustic abnormalities.

For visual saliency detection, the situation was much better due to the multitude of visual saliency models. But, many computational models were too complex, requiring several seconds if not minutes to process a single video frame. We built on the work by Hou et al. [HZ07, HHK12] and derived quaternion-based models that are state-of-the-art in predicting human gaze patterns as well as computationally lightweight. Being able to calculate a saliency map in less than one millisecond, we developed the—to our best knowledge—fastest implementation of a state-of-the-art saliency model. Having seen that the color space can substantially influence the performance of spectral saliency models, we investigated color space decorrelation as a means to provide a more appropriate image-specific color space for low-level saliency models. This way, we were able to improve the performance of several, different visual saliency algorithms.

Equipped with applicable auditory and visual saliency models, we had to address how to represent the spatial saliency distribution in the 3D environment surrounding the robot. This was an essential aspect, because it would form the foundation for audio-visual saliency fusion and subsequent aspects such as, e.g., implementing inhibition of return. Given our previous experience [SRP+09], we discarded the commonly found grid-like representations and tested a parametric Gaussian 3D model that implements the idea of salient 3D proto-objects. This way, we can represent the spatial saliency distribution as a mixture of Gaussians, independent of the modality. This, of course, makes clustering and subsequent crossmodal fusion relatively easy to implement and computationally efficient. However, it is necessary to being able to efficiently transfer the visual and auditory saliency information into such a 3D proto-object model. While we simply adapted sound source localization toward auditory proto-objects, we proposed a novel method to efficiently detect and extract salient visual proto-objects based on the isophote curvature of the saliency map.

Being able to efficiently handle audio-visual saliency information in the 3D proto-object model, we could implement the actual scene exploration. This way, we could devise a balanced approach that combines the best aspects of two strategies that we encountered in the literature, i.e. try to minimize the ego-motion and investigate the most salient regions first. Furthermore, using the two strategies as baselines for good and bad behavior, we could derive evaluation measures to quantify the quality and tradeoffs made by the balanced approach.

Future Work

There remain many interesting directions for future work. With respect to audio-visual saliency detection, first, we see a lot of potential for better bottom-up as well as top-down auditory saliency models. An interesting development in this direction is the link between pupil dilation and auditory attention [WBM12], which might allow a quantitative evaluation methodology of bottom-up auditory saliency models that is not application oriented. In contrast, visual saliency detection is a very mature field, but there seems to be room for improvement when working with videos instead of images. Additionally, due to the rise of low-cost depth cameras such as, e.g., Kinect the integration of depth information into visual saliency models is more important than ever before. It would also be very interesting to integrate high-level attentional modulation to incorporate task-based influences during, for example, visual search or human-robot interaction. However, we would like to note that all these aspects can be integrated seamlessly into our framework by adapting or replacing the auditory and visual attention models. With respect to our multiobjective exploration and the robot's overt attention, we think that it would be very interesting to investigate and evaluate how the generated head motion patterns can be made as human-like as possible.

References

[AS10]   Achanta, R., Süsstrunk, S.: Saliency detection using maximum symmetric surround. In: Proceedings of the International Conference on Image Processing (2010)

[AHES09]   Achanta, R., Hemami, S., Estrada, F., Süsstrunk, S.: Frequency-tuned salient region detection. In: Proceedings of the International Conference on Computer Vision and Pattern Recognition (2009)

[All96]   Alley, R.E.: Algorithm Theoretical Basis Document for Decorrelation Stretch. NASA, JPL (1996)

[AS13]   Alsam, A., Sharma, P.: A robust metric for the evaluation of visual saliency algorithms. J. Opt. Soc. Am. (2013)

[ARA+06]   Asfour, T., Regenstein, K., Azad, P., Schröder, J., Bierbaum, A., Vahrenkamp, N., Dillmann, R.: ARMAR-III: an integrated humanoid platform for sensory-motor control. In: Humanoids (2006)

[AWA+08]   Asfour, T., Welke, K., Azad, P., Ude, A., Dillmann, R.: The Karlsruhe Humanoid Head. In: Humanoids (2008)

[AHW+10]   Andreopoulos, A., Hasler, S., Wersing, H., Janssen, H., Tsotsos, J., Körner, E.: Active 3D object localization using a humanoid robot. IEEE Trans. Robot. 47–64 (2010)

[Bar61]   Barlow, H.: Possible principles underlying the transformation of sensory messages. Sens. Commun. 217–234 (1961)

[BS97]   Bell, A.J., Sejnowski, T.J.: The independent components of scenes are edge filters. Vis. Res. **37**(23), 3327–3338 (1997)

[BKMG10]   Begum, M., Karray, F., Mann, G.K.I., Gosine, R.G.: A probabilistic model of overt visual attention for cognitive robots. IEEE Trans. Syst. Man Cybern. B **40**, 1305–1318 (2010)

[Ber76]   Bernardo, J.M.: Algorithm as 103 psi(digamma function) computation. Appl. Stat. **25**, 315–317 (1976)

[BZ09]   Bian, P., Zhang, L.: Biological plausibility of spectral domain approach for spatiotemporal visual saliency. In: Proceedings of the Annual Conference on Neural Information Processing Systems (2009)

[BT09]   Bruce, N., Tsotsos, J.: Saliency, attention, and visual search: an information theoretic approach. J. Vis. **9**(3), 1–24 (2009)

[BSF11]   Brown, M., Susstrunk, S., Fua, P.: Spatio-chromatic decorrelation by shift-invariant filtering. In: CVPR Workshop (2011)

[BSI13a]   Borji, A., Sihite, D., Itti, L.: What/where to look next? modeling top-down visual attention in complex interactive environments. IEEE Trans. Syst. Man Cybern. A 99 (2013)

[BSI13b]   Borji, A., Sihite, D.N., Itti, L.: Quantitative analysis of human-model agreement in visual saliency modeling: a comparative study. IEEE Trans. Image Process. **22**(1), 55–69 (2013)

[BG83]   Buchsbaum, G., Gottschalk, A.: Trichromacy, opponent colours coding and optimum colour information transmission in the retina. In: Proceedings of the Royal Society, vol. B, no. 220, pp. 89–113 (1983)

[BZCM08]   Butko, N., Zhang, L., Cottrell, G., Movellan, J.R.: Visual saliency model for robot cameras. In: Proceedings of the International Conference on Robotics and Automation (2008)

[CC03]   Cashon, C., Cohen, L.: The construction, deconstruction, and reconstruction of infant face perception. NOVA Science Publishers: ch, pp. 55–68. The development of face processing in infancy and early childhood, Current perspectives (2003)

[CHEK07]   Cerf, M., Harel, J., Einhäuser, W., Koch, C.: Predicting human gaze using low-level saliency combined with face detection. In: Proceedings of the Annual Conference on Neural Information Processing Systems (2007)

[CFK08]   Cerf, M., Frady, P., Koch, C.: Subjects' inability to avoid looking at faces suggests bottom-up attention allocation mechanism for faces. In: Proceedings of the Society for Neuroscience (2008)

[CFK09]   Cerf, M., Frady, E.P., Koch, C.: Faces and text attract gaze independent of the task: experimental data and computer model. J. Vis. **9** (2009)

[CLE]   CLEAR2007: Classification of events, activities and relationships evaluation and workshop. http://www.clear-evaluation.org

[CM02]   Comaniciu, D., Meer, P.: Mean shift: a robust approach toward feature space analysis. IEEE Trans. Pattern Anal. Mach. Intell. 603–619 (2002)

[CLR90]   Cormen, T.H., Leiserson, C.E., Rivest, R.L.: Introduction to Algorithms. MIT Press and McGraw-Hill (1990)

[Cox64]   Cox, R.T.: Probability, frequency, and reasonable expectation. Am. J. Phys. **14**, 1–13 (1964)

[DBZ07]   Dankers, A., Barnes, N., Zelinsky, A.: A reactive vision system: active-dynamic saliency. In: Proceedings of the International Conference on Computer Vision Systems (2007)

[DSB01]   DiBiase, J.H., Silverman, H.F., Brandstein, M.S.: Robust localization in reverberant rooms, ch. 8, pp. 157–180. Springer (2001)

[DSMS02]   Dragoi, V., Sharma, J., Miller, E.K., Sur, M.: Dynamics of neuronal sensitivity in visual cortex and local feature discrimination. Nat. Neurosci. 883–891 (2002)

[DWM+11]   Duan, L., Wu, C., Miao, J., Qing, L., Fu, Y.: Visual saliency detection by spatially weighted dissimilarity. In: Proceedings of the Interantional Conference on Computer Vision and Pattern Recognition (2011)

[Dun84]   Duncan, J.: Selective attention and the organization of visual information. J. Exp. Psychol.: General **113**(4), 501–517 (1984)

[Ell93]   Ell, T.: Quaternion-fourier transforms for analysis of two-dimensional linear time-invariant partial differential systems. In: International Conference Decision and Control (1993)

[ES07]   Ell, T., Sangwine, S.: Hypercomplex fourier transforms of color images. IEEE Trans. Image Process. **16**(1), 22–35 (2007)

[EDR94]   Egly, R., Driver, J., Rafal, R.D.: Shifting visual attention between objects and locations: evidence from normal and parietal lesion subjects. J. Exp. Psychol.: General, **123**(2) (1994)

[Ehr05]   Ehrgott, M.: Multicriteria Optimization. Springer (2005)

[ESJ86]   Eriksen, C.W.: St James, J.D.: Visual attention within and around the field of focal attention: a zoom lens model. Percept. Psychophys. **40**(4), 225–240 (1986)

[Ess00]   Essa, I.: Ubiquitous sensing for smart and aware environments. IEEE Pers. Commun. **7**(5), 47–49 (2000)

[FPB06]   Fleming, K.A., Peters II, R.A., Bodenheimer, R.E.: Image mapping and visual attention on a sensory ego-sphere In: Proceedings of the International Conference on Intelligent Robotics and Systems (2006)

[FH08]   Feng, W., Hu, B.: Quaternion discrete cosine transform and its application in color template matching. In: International Conference on Image and Signal Processing, pp. 252–256 (2008)

[FRC10]   Frintrop, S., Rome, E., Christensen, H.I.: Computational visual attention systems and their cognitive foundation: a survey. ACM Trans. Appl. Percept. **7**(1), 6:1–6:39 (2010)

[FE04]   Fröba, B., Ernst, A.: Face detection with the modified census transform. In: Proceedings of the International Conference on Automatic Face and Gesture Recognition (2004)

[GMV08]   Gao, D., Mahadevan, V., Vasconcelos, N.: On the plausibility of the discriminant center-surround hypothesis for visual saliency. J. Vis. **8**(7), 1–18 (2008)

[GvdBSG01]   Geusebroek, J.M., van den Boomgaard, R., Smeulders, A.W.M., Geerts, H.: Color invariance. IEEE Trans. Pattern Anal. Mach. Intell. **23**(12), 1338–1350 (2001)

[GSvdW03]   Geusebroek, J.-M., Smeulders, A., van de Weijer, J.: Fast anisotropic gauss filtering. IEEE Trans. Image Process. **12**(8), 938–943 (2003)
[GKW87]   Gillespie, A.R., Kahle, A.B., Walker, R.E.: Color enhancement of highly correlated images. II. Channel ratio and chromaticity transformation techniques. Remote Sens. Environ. **22**(3), 343–365 (1987)
[Gil00]   Gillies, D.: The subjective theory. In: Philosophical Theories of Probability. Routledge, ch. 4 (2000)
[GZMT12]   Goferman, S., Zelnik-Manor, L., Tal, A.: Context-aware saliency detection. IEEE Trans. Pattern Anal. Mach, Intell (2012)
[GZ10]   Guo, C., Zhang, L.: A novel multiresolution spatiotemporal saliency detection model and its applications in image and video compression. IEEE Trans. Image Process. **19**, 185–198 (2010)
[GMZ08]   Guo, C., Ma, Q., Zhang, L.: Spatio-temporal saliency detection using phase spectrum of quaternion fourier transform. In: Proceedings of the International Conference on Computer Vision and Pattern Recognition (2008)
[HL08]   Hall, D., Linas, J.: Handbook of Multisensor Data Fusion: Theory and Practice. CRC Press (2008)
[Ham66]   Hamilton, W.R.: Elements of Quaternions. University of Dublin Press (1866)
[HKP07]   Harel, J., Koch, C., Perona, P.: Graph-based visual saliency. In: Proceedings of the Annual Conference on Neural Information Processing Systems (2007)
[HB95]   Heeger, D.J., Bergen, J.R.: Pyramid-based texture analysis/synthesis. In: Proceedings of the Techniques Annual Conference Special Interest Group on Graphics and Interactive, pp. 229–238 (1995)
[Hen03]   Henderson, J.M.: Human gaze control during real-world scene perception. Trends Cogn. Sci. 498–504 (2003)
[HKM+09]   Heracles, M., Körner, U., Michalke, T., Sagerer, G., Fritsch, J., Goerick, C.: A dynamic attention system that reorients to unexpected motion in real-world traffic environments. In: Proceedings of the International Conference on. Intelligent Robots and Systems (2009)
[Her64]   Hering, E.: Outlines of a Theory of the Light Sense. Harvard University Press (1964)
[HO07]   Hershey, J., Olsen, P.: Approximating the kullback leibler divergence between gaussian mixture models. In: Proceedings of the International Conference on Acoustics, Speech, and Signal Processing (2007)
[HA04]   Hodge, V.J., Austin, J.: A survey of outlier detection methodologies. Artif. Intell. Rev. **22**, 85–126 (2004)
[HY09]   Holsopple, J., Yang, S.: Designing a data fusion system using a top-down approach. In: Proceedings of the International Conference for Military Communications (2009)
[HZ07]   Hou, X., Zhang, L.: Saliency detection: a spectral residual approach. In: Proceedings of the International Conference on Computer Vision and Pattern Recognition (2007)
[HHK12]   Hou, X., Harel, J., Koch, C.: Image signature: highlighting sparse salient regions. IEEE Trans. Pattern Anal. Mach. Intell. **34**(1), 194–201 (2012)
[HBD75]   Huang, T., Burnett, J., Deczky, A.: The importance of phase in image processing filters. IEEE Trans. Acoust. Speech Signal Process. **23**(6), 529–542 (1975)
[IB09]   Itti, L., Baldi, P.: Bayesian surprise attracts human attention. Vis. Res. **49**(10), 1295–1306 (2009)
[IB05]   Itti, L., Baldi, P.F.: A principled approach to detecting surprising events in video. In: Proceedings of the International Conference on Image Processing Computer Vision and Pattern Recognition (2005)
[IB06]   Itti, L., Baldi, P.F.: Bayesian surprise attracts human attention. In: Proceedings of the Annual Conference on Neural Information Processing Systems (2006)
[IK00]   Itti, L., Koch, C.: A saliency-based search mechanism for overt and covert shifts of visual attention. Vis. Res. **40**(10–12), 1489–1506 (2000)
[IKN98]   Itti, L., Koch, C., Niebur, E.: A model of saliency-based visual attention for rapid scene analysis. IEEE Trans. Pattern Anal. Mach. Intell. **20**(11), 1254–1259 (1998)

[Jay03]   Jaynes, E.T.: Probability Theory. The Logic of Science Cambridge University Press (2003)

[JEDT09]   Judd, T., Ehinger, K., Durand, F., Torralba, A.: Learning to predict where humans look. In: Proceedings of the International Conference on Computer Vision (2009)

[JDT11]   Judd, T., Durand, F., Torralba, A.: Fixations on low-resolution images. J. Vis. **11**(4) (2011)

[JDT12]   Judd, T., Durand, F., Torralba, A.: A benchmark of computational models of saliency to predict human fixations. Technical Report, MIT (2012)

[JM97]   Johnson, D., McGeoch, L.: The traveling salesman problem: a case study in local optimization. Local search in combinatorial optimization, pp. 215–310 (1997)

[JOvW+05]   Jost, T., Ouerhani, N., von Wartburg, R., Mäuri, R., Häugli, H.: Assessing the contribution of color in visual attention. Comput. Vis. Image Underst. **100**, 107–123 (2005)

[Kal09]   Kalinli, O.: Biologically inspired auditory attention models with applications in speech and audio processing, Ph.D. dissertation, University of Southern California, Los Angeles, CA, USA (2009)

[KN09]   Kalinli, O., Narayanan, S.: Prominence detection using auditory attention cues and task-dependent high level information. IEEE Trans. Audio Speech Lang. Proc. **17**(5), 1009–1024 (2009)

[KT00]   Kahneman, D., Treisman, A.: Varieties of Attention. Academic Press (2000), ch. Changing views of attention and automaticity, pp. 26–61

[KTG92]   Kahneman, D., Treisman, A., Gibbs, B.J.: The reviewing of object files: object-specific integration of information. Cogn. Psychol. **24**(2), 175–219 (1992)

[KPLL05]   Kayser, C., Petkov, C.I., Lippert, M., Logothetis, N.K.: Mechanisms for allocating auditory attention: an auditory saliency map. Curr. Biol. **15**(21), 1943–1947 (2005)

[KJS+02]   Klin, A., Jones, W., Schultz, R., Volkmar, F., Cohen, D.: Visual fixation patterns during viewing of naturalistic social situations as predictors of social competence in individuals with autism. Arch. Gen. Psychiatry **59**(9), 809–816 (2002)

[KNd08]   Kootstra, G., Nederveen, A., de Boer, B.: Paying attention to symmetry. In: Proceedings of the British Conference on Computer Vision (2008)

[KBS+10]   Kühn, B., Belkin, A., Swerdlow, A., Machmer, T., Beyerer, J., Kroschel, K.: Knowledge-driven opto-acoustic scene analysis based on an object-oriented world modelling approach for humanoid robots. In: Proceedings of the 41st International Symposium Robotics and 6th German Conference on Robotics (2010)

[LLAH11]   Li, J., Levine, M.D., An, X., He, H.: Saliency detection based on frequency and spatial domain analysis. In: Proceedings of the British Conference on Computer Vision (2011)

[LSL00]   Liang, Y., Simoncelli, E., Lei, Z.: Color channels decorrelation by ica transformation in the wavelet domain for color texture analysis and synthesis. Proceedings of the International Conference on Computer Vision and Pattern Recognition **1**, 606–611 (2000)

[LHR05]   Lichtenauer, J., Hendriks, E. Reinders, M.: Isophote properties as features for object detection. In: Proceedings of the International Conference on Computer Vision and Pattern Recognition (2005)

[LZG+12]   Lin, K.-H., Zhuang, X., Goudeseune, C., King, S., Hasegawa-Johnson, M., Huang, T.S.: Improving faster-than-real-time human acoustic event detection by saliency-maximized audio visualization. In: Proceedings of the International Conference on Acoustics, Speech, and Signal Processing (2012)

[LL12]   Lu, S., Lim, J.-H.: Saliency modeling from image histograms. In: Proceedings of the European Conference on Computer Vision (2012)

[LLLN12]   Luo, W., Li, H., Liu, G., Ngan, K.N.: Global salient information maximization for saliency detection. Signal Process.: Image Commun. **27**, 238–248 (2012)

[MMS+09]   Machmer, T., Moragues, J., Swerdlow, A., Vergara, L., Gosalbez-Castillo, J., Kroschel, K.: Robust impulsive sound source localization by means of an energy detector for temporal alignment and pre-classification. In: Proceedings of the European Signal Processing of Conference (2009)

[MSKK10]   Machmer, T., Swerdlow, A., Kühn, B., Kroschel, K.: Hierarchical, knowledge-oriented opto-acoustic scene analysis for humanoid robots and man-machine interaction. In: Proceedings of the International Conference on Robotics and Automation (2010)

[MFL+07]   Meger, D., Forssén, P.-E., Lai, K., Helmer, S., McCann, S., Southey, T., Baumann, M., Little, J.J., Lowe, D.G.: Curious George: an attentive semantic robot. In: IROS Workshop: From sensors to human spatial concepts (2007)

[MCB06]   Meur, O.L., Callet, P.L., Barba, D.: Predicting visual fixations on video based on low-level visual features. J. Vis. 47(19), 2483–2498 (2006)

[MMKL99]   Muller, J.R., Metha, A.B., Krauskopf, J., Lennie, P.: Rapid adaptation in visual cortex to the structure of images. Science 285, 1405–1408 (1999)

[NSK14]   Nakajima, J., Sugimoto, A., Kawamoto, K.: Incorporating audio signals into constructing a visual saliency map. In: Klette, R., Rivera, M., Satoh, S. (eds.) Image and Video Technology, Series Lecture Notes in Computer Science, vol. 8333. Springer, Berlin, Heidelberg (2014)

[OK04]   Olmos, A., Kingdom, F.A.A.: A biologically inspired algorithm for the recovery of shading and reflectance images. Perception 33, 1463–1473 (2004)

[OF96]   Olshausen, B.A., Field, D.J.: Emergence of simple-cell receptive field properties by learning a sparse code for natural images. Nature 381, 607–609 (1996)

[OLK07]   Onat, S., Libertus, K., König, P.: Integrating audiovisual information for the control of overt attention. J. Vis. 7(10) (2007)

[OL81]   Oppenheim, A., Lim, J.: The importance of phase in signals. Proc. IEEE 69(5), 529–541 (1981)

[OMS08]   Orabona, F., Metta, G., Sandini, G.: A proto-object based visual attention model. In: Paletta, L., Rome, E. (eds.) Attention in Cognitive Systems. Theories and Systems from an Interdisciplinary Viewpoint, pp. 198–215 (2008)

[PLN02]   Parkhurst, D., Law, K., Niebur, E.: Modeling the role of salience in the allocation of overt visual attention. Vis. Res. 42(1), 107–123 (2002)

[Pas08]   Pascale, D.: A review of RGB color spaces...from xyY to R'G'B' (2008)

[PI08a]   Peters, R.J., Itti, L.: Applying computational tools to predict gaze direction in interactive visual environments. ACM Trans. Appl. Percept. 5(2) (2008)

[PI08b]   Peters, R., Itti, L.: The role of fourier phase information in predicting saliency. J. Vis. 8(6), 879 (2008)

[PIIK05]   Peters, R., Iyer, A., Itti, L., Koch, C.: Components of bottom-up gaze allocation in natural images. Vis. Res. 45(18), 2397–2416 (2005)

[Pos80]   Posner, M.I.: Orienting of attention. Q. J. Exp. Psychol. 32(1), 3–25 (1980)

[RBC06]   Rajashekar, U., Bovik, A.C., Cormack, L.K.: Visual search in noise: revealing the influence of structural cues by gaze-contingent classïňǍcation image analysis. J. Vis. 6(4), 379–386 (2006)

[RMDB+13] Ramenahalli, S., Mendat, D.R., Dura-Bernal, S., Culurciello, E., Niebur, E., Andreou, A.: Audio-visual saliency map: overview, basic models and hardware implementation. In: Annual Conference on Information Sciences and Systems (2013)

[RB99]   Rao, R.P., Ballard, D.H.: Predictive coding in the visual cortex: a functional interpretation of some extra-classical receptive-field effects. Nat. Neurosci. 79–87 (1999)

[Rat65]   Ratliff, F.: Mach Bands: Quantitative Studies on Neural Networks in the Retina. Holden-Day, San Francisco (1965)

[RP11]   Reinhard, E., Pouli, T.: Colour spaces for colour transfer. Computational Color Imaging, series Lecture Notes in Computer Science 6626, 1–15 (2011)

[RAGS01]   Reinhard, E., Adhikhmin, M., Gooch, B., Shirley, P.: Color transfer between images. IEEE Comput. Graph. Appl. 21(5), 34–41 (2001)

[Ren00a]   Rensink, R.A.: The dynamic representation of scenes. Vis. Cogn. 7, 17–42 (2000)

[Ren00b]   Rensink, R.A.: Seeing, sensing, and scrutinizing. Vis. Res. **40**, 1469–1487 (2000)
[RDM+13]   Riche, N., Duvinage, M., Mancas, M., Gosselin, B., Dutoit, T.: Saliency and human
            fixations: state-of-the-art and study of comparison metrics. In: Proceedings of the
            International Conference on Computer Vision (2013)
[Rob]   RobotCub Consortium: iCub—an open source cognitive humanoid robotic platform.
            http://www.icub.org
[RCC98]   Ruderman, D., Cronin, T., Chiao, C.: Statistics of cone responses to natural images:
            implications for visual coding. J. Opt. Soc. Am. **15**(8), 2036–2045 (1998)
[RLB+08]   Ruesch, J., Lopes, M., Bernardino, A., Hornstein, J., Santos-Victor, J., Pfeifer, R.:
            Multimodal saliency-based bottom-up attention: a framework for the humanoid
            robot iCub. In: Proceedings of the International Conference on Robotics and
            Automation (2008)
[RLS98]   Roelfsema, P.R., Lamme, V.A.F., Spekreijse, H.: Object-based attention in the pri-
            mary visual cortex of the macaque monkey. Nature **395**, 376–381 (1998)
[San96]   Sangwine, S.J.: Fourier transforms of colour images using quaternion or hypercom-
            plex, numbers. Electron. Lett. **32**(21), 1979–1980 (1996)
[SE00]   Sangwine, S., Ell, T.: Colour image filters based on hypercomplex convolution. IEEE
            Proc. Vis. Image Signal Process. **147**(2), 89–93 (2000)
[SSYK07]   Saidi, F., Stasse, O., Yokoi, K., Kanehiro, F.: Online object search with a humanoid
            robot. In: Proceedings of the International Conference on Intelligent Robots and
            Systems (2007)
[SRP+09]   Schauerte, B., Richarz, J., Plötz, T., Thurau, C., Fink, G.A.: Multi-modal and multi-
            camera attention in smart environments. In: Proceedings of the 11th International
            Conference on Multimodal Interfaces (ICMI). ACM, Cambridge, MA, USA, Nov
            2009
[SRF10]   Schauerte, B., Richarz, J., Fink, G.A.: Saliency-based identification and recogni-
            tion of pointed-at objects. In: Proceedings of the 23rd International Conference on
            Intelligent Robots and Systems (IROS). IEEE/RSJ, Taipei, Taiwan, Oct. 2010
[SF10a]   Schauerte, B., Fink, G.A.: Focusing computational visual attention in multi-modal
            human-robot interaction. In: Proceedings of the 12th International Conference on
            Multimodal Interfaces and 7th Workshop on Machine Learning for Multimodal
            Interaction (ICMI-MLMI). ACM, Beijing, China, Nov. 2010
[SIK11]   Schnupp, J., Nelken, I., King, A.: Auditory Neuroscience. MIT Press (2011)
[SY06]   Serences, J.T., Yantis, S.: Selective visual attention and perceptual coherence. Trends
            Cogn. Sci. **10**(1), 38–45 (2006)
[SS07]   Shic, F., Scassellati, B.: A behavioral analysis of computational models of visual
            attention. Int. J. Comput. Vis. **73**, 159–177 (2007)
[SW87]   Shulman, G.L., Wilson, J.: Spatial frequency and selective attention to spatial loca-
            tion. Perception **16**(1), 103–111 (1987)
[SS06]   Simion, C., Shimojo, S.: Early interactions between orienting, visual sampling and
            decision making in facial preference. Vis. Res. **46**(20), 3331–3335 (2006)
[SG31]   Smith, T., Guild, J.: The C.I.E. colorimetric standards and their use. Trans. Opt. Soc.
            33(3), 73 (1931)
[SPG12]   Song, G., Pellerin, D., Granjon, L.: How different kinds of sound in videos can influ-
            ence gaze. In: Interantional Workshop on Image Analysis for Multimedia Interactive
            Services (2012)
[TBG05]   Tatler, B., Baddeley, R., Gilchrist, I.: Visual correlates of fixation selection: effects
            of scale and time. J. Vis. **45**(5), 643–659 (2005)
[TMZ+07]   Temko, A., Malkin, R., Zieger, C., Macho, D., Nadeu, C., Omologo, M.: Clear
            evaluation of acoustic event detection and classification systems. In: Stiefelhagen,
            R., Garofolo, J. (eds.) Series Lecture Notes in Computer Science, vol. 4122, pp.
            311–322. Springer, Berlin, Heidelberg (2007)
[TDW91]   Tipper, S.P., Driver, J., Weaver, B.: Object-centred inhibition of return of visual
            attention. Q. J. Exp. Psychol. **43**, 289–298 (1991)

[TOCH06]   Torralba, A., Oliva, A., Castelhano, M.S., Henderson, J.M.: Contextual guidance of eye movements and attention in real-world scenes: the role of global features in object search. Psychol. Rev. 113(4) (2006)

[TG80]   Treisman, A.M., Gelade, G.: A feature-integration theory of attention. Cogn. Psychol. 12(1), 97–136 (1980)

[Tso89]   Tsotsos, J.K.: The complexity of perceptual search tasks. In: Proceedings of the International Joint Conference on Artificial Intelligence (1989)

[Tso95]   Tsotsos, J.K.: Behaviorist intelligence and the scaling problem. Artif. Intell. 75, 135–160 (1995)

[Tso11]   Tsotsos, J.K.: A Computational Perspective on Visual Attention. The MIT Press (2011)

[vR79]   van Rijsbergen, C.J.: Information Retrieval, 2nd edn. Butterworth (1979)

[VCSS01]   Vijayakumar, S., Conradt, J., Shibata, T., Schaal, S.: Overt visual attention for a humanoid robot. In: Proceedings of the International Conference on Intelligent Robotics and Systems (2001)

[WK06]   Walther, D., Koch, C.: Modeling attention to salient proto-objects. Neural Networks 19(9), 1395–1407 (2006)

[WBM12]   Wang, C.-A., Boehnke, S., Munoz, D.: Pupil dilation evoked by a salient auditory stimulus facilitates saccade reaction times to a visual stimulus. J. Vis. 12(9), 1254 (2012)

[Wel11]   Welke, K.: Memory-based active visual search for humanoid robots, Ph.D. dissertation, Karlsruhe Institute of Technology (2011)

[WAD09]   Welke, K., Asfour, T., Dillmann, R..: Active multi-view object search on a humanoid head. In: Proceedings of the International Conference on Robotics and Automation (2009)

[WAD11]   Welke, K., Asfour, T., Dillmann, R.: Inhibition of return in the bayesian strategy to active visual search. In: Proceedings of the International Conference on Machine Vision Applications (2011)

[Weg05]   Wegener, I.: Theoretische Informatik—eine algorithmenorientierte Einführung. Teubner (2005)

[Wikc]   Wikimedia Common (Googolplexbyte): Diagram of the opponent process. http://commons.wikimedia.org/wiki/File:Diagram_of_the_opponent_process.png, retrieved 3 April 2014, License CC BY-SA 3.0

[WS13]   Winkler, S., Subramanian, R.: Overview of eye tracking datasets. In: International Workshop on Quality of Multimedia Experience (2013)

[WCD+13]   Wu, P.-H., Chen, C.-C., Ding, J.-J., Hsu, C.-Y., Huang, Y.-W.: Salient region detection improved by principle component analysis and boundary information. IEEE Trans. Image Process. 22(9), 3614–3624 (2013)

[XCKB09]   Xu, T., Chenkov, N., Kühnlenz, K., Buss, M.: Autonomous switching of top-down and bottom-up attention selection for vision guided mobile robots. In: Proceedings of the International Conference on Intelligent Robotics and Systems (2009)

[XPKB09]   Xu, T., Pototschnig, T., Kühnlenz, K., Buss, M.: A high-speed multi-GPU implementation of bottom-up attention using CUDA. In: Proceedings of the International Conference on Robotics and Automation (2009)

[YGMG13]   Yu, Y., Gu, J., Mann, G., Gosine, R.: Development and evaluation of object-based visual attention for automatic perception of robots. IEEE Trans. Autom. Sci. Eng. 10(2), 365–379 (2013)

[Zad65]   Zadeh, L.: Fuzzy sets. Inform. Control 8(3), 338–353 (1965)

[ZK11]   Zhao, Q., Koch, C.: Learning a saliency map using fixated locations in natural scenes. J. Vis. 11(3), 1–15 (2011)

[ZTM+08]   Zhang, L., Tong, M.H., Marks, T.K., Shan, H., Cottrell, G.W.: Sun: a bayesian framework for saliency using natural statistics. J. Vis. 8(7) (2008)

[ZJY12]   Zhou, J., Jin, Z., Yang, J.: Multiscale saliency detection using principle component analysis. In: International Joint Conference on Neural Networks, pp. 1–6 (2012)

Chapter 4
Multimodal Attention with Top-Down Guidance

In many situations, people want to guide our attention to specific objects or aspects in our environment. In fact, we have already seen an example of such attentional guidance as being part of advertisement design, see Fig. 4.1 or 2.10. However, such attentional guidance is not just a factor in effective advertisement. Instead, it is a natural process and part of everyday natural communication that we are not just frequently subjected to but often exercise ourselves—consciously as well as unconsciously. For example, when persons interact, interpreting non-verbal attentional signals such as, most importantly, pointing gestures [LB05] and gaze [BDT08] are essential to establish a joint focus of attention—i.e., a common understanding of what we are talking and thinking about. Human infants develop the ability to interpret related non-verbal signals around the age of 1 year. This is a very important step in infant development, because it enables infants to associate verbal descriptions with the visual appearances of objects [Tom03, KH06]. This ability provides the means to acquire a common verbal dictionary and enable verbal communication with other humans, which is important to build strong social connections. As a consequence, there exists evidence "that joint attention reflects mental and behavioral processes in human learning and development" [MN07b].

In this chapter, we want to determine where other people want us to look at. In other words, we want to answer the question: Which object forms the intended focus of the current scene or situation? We address two different domains: First, we investigate how we can model attentional guidance in human-robot interaction. During human-human and consequently human-robot interaction, we frequently use non-verbal (e.g., pointing gestures, head nods, or eye gaze) and verbal (e.g., object descriptions) to guide the attention of our conversation partner toward specific objects in the environment. Here, we focus on spoken object descriptions and pointing gestures. Second, we want to determine the object of interest in web images—in principle, in all forms of visual media that is created by human photographers, camera men, directors, etc. In such images and videos, it is not just the visible content such as, for example, the gaze of other people (see Fig. 4.1) that influences where we look, but the whole image is composed in such a way that the most important, relevant, and

© Springer International Publishing Switzerland 2016
B. Schauerte, *Multimodal Computational Attention for Scene Understanding and Robotics*, Cognitive Systems Monographs 30,
DOI 10.1007/978-3-319-33796-8_4

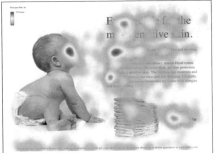

Fig. 4.1 Knowing how top-down cues can guide attention is an important aspect of advertisement design. Images from and used with permission from James Breeze

thus interesting object is highlighted. For example, photographer's often place the object of interest in the image's center or follow guidelines such as the rule of thirds to compose images.

In both domains, in contrast to our approach to evaluate visual saliency in Sect. 3.2, we do not try to predict where people will look. Instead, all images depict a specific object of interest and we try to determine this object. Therefore, we estimate a target object area and/or a sequence of target locations that are highly likely to be part of the intended object. This way, if we wanted to recognize the object, we only have to run our object classifiers on a very limited part of the image, thus saving computational resources. Furthermore, we can direct a robot's sensors—i.e., active vision—to focus on each of these target object hypotheses until we have seen the right object, which is similar to our approach in Sect. 3.4, but now includes top-down target information that we can acquire through interaction. If we want to learn—e.g., from weakly labeled web images or through passive observation of human behavior—, then the limited set of image locations can serve as a prior to train better models.[1]

Remainder

Complementary to our broad background discussion in Chap. 2, we provide an overview of related work that is relevant to understand the contributions presented in this chapter (Sect. 4.1). Then, we present how we adapt a state-of-the-art salient object detection algorithm to remove its implicit center bias (Sect. 4.2). Afterwards, we present how we can determine the target object in the presence of pointing gestures and linguistic descriptions of primitive visual attributes (Sect. 4.3). Finally, we conclude the technical part of this book and present how we can use the introduced methods to determine objects that are being looked at in images that we collected from Flickr (Sect. 4.4).

---

[1] We successfully implemented this idea to learn robust color term models from web images [SS12a]. For this purpose, we used salient object detection to serve as a spatial prior to weight the information at each image location.

4.1 Related Work and Contributions

In the following, we first present the most important related work for each of the affected research topics. Then, after each topic's overview, we discuss our contribution with respect to the state-of-the-art.

4.1.1 Joint Attention

To establish a joint focus of attention describes the human ability to verbally and non-verbally coordinate the focus of attention with interaction partners (see our introduction to this chapter). On one side this is achieved by directing the attention toward interesting objects, persons, or events, and on the other side by responding to these attention directing signals. Since this ability is one of the most important aspects of natural communication and social interaction, it has been addressed in various research areas such as, most importantly: psychology (e.g., [Ban04, LB05, MN07b]), computational linguistics (e.g., [SC09]), and robotics (e.g., [Bro07, KH06, NHMA03, SC09, SKI+07, FBH+08, YSS10]). Especially for social robots the ability to initiate (e.g., [DSS06, SC09, SKI+07]) and respond to (e.g., [Bro07, SKI+07, TTDC06]) signals related to achieve joint attention are crucial aspects of natural and human-like interaction.

In the following, we address two specific aspects of responding to joint attention signals, i.e. how verbal object descriptions and pointing gestures can influence attention and guide visual search. In this context, we also address the influence of gaze, although we do not integrate or evaluate gaze as a feature in a human-robot interaction (HRI) domain.

4.1.1.1 Pointing

Pointing gestures are an important non-verbal signal to direct the attention toward a spatial region or direction and establish a joint focus of attention (cf. [Ban04, GRK07, HSK+10, LB05]). Accordingly, visually recognizing pointing gestures and inferring a referent or target direction has been addressed by several authors; e.g., for interaction with smart environments (e.g., [RPF08]), wearable visual interfaces (e.g., [HRB+04]), and robots (e.g., [HSK+10, KLP+06, NS07, SYH07, SHH+08, DSHB11]). Nickel and Stiefelhagen evaluated three different methods to estimate the indicated direction of a pointing gesture to determine the referent [NS07]. Most importantly, they achieved the best target object identification results with the line-of-sight model, i.e. the pointing ray originates from the hand and follows the head-hand direction. Interestingly, this model is even sometimes used in psychological literature (e.g., [BO06]), where it has been found that the inherent inaccuracy of pointing gestures suggests that "pointing shifts attention into the visual periphery, rather than identifying referents" directly [Ban04]. Unfortunately, almost all technical systems require that the objects present in the scene are already detected, segmented, recognized, categorized and/or their attributes identified (e.g., [NS07, DSHB11]). For example, Droeschel et al. define that the pointed-at object is the object with the minimum distance to the pointing vector [DSHB11]. We would like to note at this point that we could simply implement Droeschel et al.'s approach even for unknown objects based on the proto-object model that we use for scene exploration in Sects. 3.4 and 3.5.

In principle, non-verbal signals such as pointing gestures circumscribe a referential domain to direct the attention toward an approximate spatial region (see [Ban04]). Naturally, this can clearly identify the referent in simple, non-ambiguous situations. However, as pointing gestures are inherently inaccurate in ambiguous situations (see [BI00, KLP+06]), context knowledge may be necessary to clearly identify the referent (see [LB05, SKI+07, SF10a]).

4.1.1.2 Language

Language can provide contextual knowledge about the referent such as, e.g., spatial relations and information about the object's visual appearance (see, e.g., [KCP+13, SF10a]). Most importantly, verbal and non-verbal references can be seen to form composite signals, i.e. the speaker will compensate the inaccuracy or ambiguity of one signal with the other (see [Ban04, BC98, GRK07, KLP+06, LB05, Piw07, SKI+07]). Even without additional non-verbal signals, what we see in the environment or in a scene is often necessary to resolve ambiguities in otherwise ambiguous English sentences (e.g., [KC06, HRM11]), while at the same time such sentences influence where we look in scenes, i.e. our gaze patterns.

When directly verbally referring to an object, most information about the referent is encoded in noun-phrases (see, e.g., [MON08]), which consist of determiners (e.g., "that"), modifiers (e.g., "red") and a head-noun (e.g., "book"). To analyze the structure of sentences and extract such information, tagging and shallow parsing can be applied. In corpus linguistics, part-of-speech (POS) tagging marks the words of a sentence with their grammatical function, e.g., demonstrative, adjective, and noun. Based on these grammatical tags and the original sentence, shallow parsing determines the constituents of a sentence as, e.g., noun-phrases. Commonly, machine learning methods are used to train taggers and shallow parsers on manually tagged linguistic corpora (e.g., [Fra79, TB00]; cf. [Bri95]). The well-established Brill tagger uses a combination of defined and learned transformation rules for tagging [Bri95]. To apply the transformation rules it requires an initial tagging, which can be provided by stochastic n-gram or regular expression taggers (cf. [Bri95]).

4.1.1.3 Color Terms

When verbally referring-to objects, relative and absolute features can be used to describe the referent (cf. [BC98]). Relative features require reference entities for identification (e.g., "the left cup", or "the big cup"), whereas absolute features do not require comparative object entities (e.g., "the red cup"). Possibly the most fundamental absolute properties of an object are its name, class, and color. When verbally referring-to color, color terms (e.g., "green", "dark blue", or "yellow-green") are used to describe the perceived color (see [Moj05]). In [BK69], the cross-cultural concept of universal "basic color terms" is introduced, circumscribing that there exists a limited set of basic color terms in each language of which all other colors are considered to be variants (e.g., the 11 basic color terms for English are: "black," "white," "red," "green," "yellow," "blue," "brown," "orange," "pink," "purple," and "gray").

In order to relate the visual appearance of objects with appropriate color terms, color models for the color terms are required. Traditionally these models are either manually defined by experts or derived from collections of manually labeled color-chips (cf. [Moj05]). Alternatively, image search engines in the Internet can be used in order to collect huge weakly labeled datasets, which make it possible to use machine learning and train robust color naming models (e.g., [vdWSV07, SS12a]).

4.1.1.4   Gaze

As we have seen in the introduction, where other people look at can and most likely will influence where a human observer will look. Like pointing gestures, gaze directions are a common signal to establish reference and infants show signs of following observed gaze directions of caregivers already at an age of 6 months [Hob05], well after infants have shown their first attraction to faces [CC03]. Again, like pointing gestures, an observed gaze direction steers the attention toward an approximate spatial region, along a corridor of attention, to establish a joint focus of attention (see, e.g., [TTDC06, BDT08]).

Since where and at what people look at is an interesting information for many applications (e.g., advertisement, driver assistance, and user interfaces), gaze estimation has been an active research area for over two decades (e.g., [BP94, SYW97, TKA02, HJ10, VSG12]). But, despite all the research effort it has attracted, gaze estimation is still an unsolved problem; e.g., even today there does not exist a gaze estimation method that can reliably estimate the gaze direction or—even more interestingly—the looked at object in an unconstrained domain such as web images. Most existing approaches focus on constrained scenarios such as, e.g., limited head poses (e.g., [SMSK08]) and/or rely on more reliable but only approximate estimates such as upper body orientation or head pose (e.g., [VS08, RVES10]). Almost all approaches do not take into account the visible objects in the environment. Consequently, the estimation of a gaze direction and the subsequent deduction of the looked at object are treated as separate steps, where the latter is usually not even addressed by the authors or, similar to pointing gestures, requires that potential targets are already known.

Highly related to our work is the work by Yücel et al. [YScM+13], which combines visual saliency and an estimated gaze direction to highlight the object that is being looked at in a simple HRI scenario. Here, a person looks at objects on a table, where the objects are distributed apart from each other and have a high perceptual saliency.

Contributions

We model how verbal descriptions, pointing gestures, and visible eye gaze can influence visual attention and highlight intended target objects. We approach this topic with the intention of being able to interpret joint attention signals and identify the intended target object without or with only very limited information about the target object's visual appearance. For this purpose, our approach relies on saliency to direct the attention toward the referent. Accordingly, we use saliency as a kind of generalized object detector (see [ADF10]), which is related to the assumption that interesting objects are often visually salient [EI08]. Consequently, our work is very different to almost all work that tries to interpret pointing gestures and gaze signals (Yücel et al. [YScM+13] being a notable exception), because we do not require any a-priori knowledge about the objects in our environment. In principle, this enables us to use gestures and gaze to guide the attention and teach our system knowledge about previously unknown objects, which we demonstrated as part of the ReferAT dataset collection that will be presented in Sect. 4.3.2.5. This also means that, in contrast to most work on gaze and pointing gesture detection, recognition, and interpretation, we are not interested in a highly precise estimated gaze or pointing direction itself. Instead, our goal is to determine the image regions that most likely depict the looked-at or pointed-at target object.

Related to our scene exploration and analysis system (Sects. 3.4 and 3.5), this again means that by focusing on the most salient areas we can improve the perception through active vision—an aspect that we have demonstrated as part of our work [SRF10] and in Sect. 3.4—and at the same time reduce the amount of data that has to be processed and analyzed. Here, it is interesting that our approach almost guarantees it that the correct target object will be focused after only a few focus of attention shifts.

As a sidenote, although we do not present the details of how we learn color term models in this book, we have improved the state-of-the-art in color term learning with two innovations: First, we developed a probabilistic color model to learn color models that better reflect natural color distributions despite the fact they have been trained on web images that contain many post-processed or even entirely artificial images [SF10b]. We have shown that this way the learned models make more "human-like" errors, if they make errors. Second, we use salient object detection as a means to predict and weight the relevance and thus influence of each image pixel's color information during training [SS12a].

4.1.2 Visual Attention

In principle, visual saliency models try to predict "interesting" image regions that are likely to attract human interest and, as a consequence, gaze. We have already discussed many aspects of bottom-up visual attention in earlier sections (Sects. 2.1.1, 3.1.1, and 3.2). Consequently, we do not address bottom-up saliency models and focus on two related but different types of models: First, salient object detection methods that try to identify and segment the most important or prominent object in an image. Second, visual attention models that allow to integrate knowledge for goal-directed adaptation of the visual saliency to support visual search.

4.1.2.1 Salient Object Detection

Most generally, "salient regions" in an image are likely to grab the attention of human observers. The task of "traditional" saliency detection is to predict where human observers look when presented with a scene, which can be recorded using eye tracking equipment, see Sect. 3.2. In 2007, Liu et al. adapted the traditional definition of visual saliency by incorporating the high level concept of a salient object into the process of visual attention computation [LSZ+07]. A "salient object" is defined as being the (most prominent) object in an image that attracts most of the user's interest. Accordingly, Liu et al. [LSZ+07] defined the task of "salient object detection" as the binary labeling problem to separate the salient object from the background. Here, it is important to note that the selection of a salient object happens consciously by the user whereas the gaze trajectories that are recorded using eye trackers are the result of mostly unconscious processes. Consequently, also taking into account that salient objects attract human gaze (see, e.g., [ESP08]), salient object detection and predicting where people look are very closely related yet substantially different tasks.

In 2009, Achanta et al. [AHES09, AS10] introduced a salient object detection approach that basically relies on the difference of pixels to the average color and intensity value. To evaluate their approach, they selected a subset of 1000 images of the image dataset that was collected from the web by Liu et al. [LSZ+07] and calculated segmentation masks of the salient objects that were marked by 9 participants using (rough) rectangle annotations [LSZ+07]. Since it was created, the salient object dataset by Achanta et al. serves as reference dataset to evaluate methods for salient object detection (see, e.g., [AHES09, AS10, KF11, CZM+11]).

Since Liu et al. defined salient object detection as binary labeling (i.e., binary segmentation) problem, it comes at no surprise that Liu et al. applied conditional random fields (CRFs) to detect salient objects, because CRFs have achieved state-of-the-art performance for several segmentation tasks such as, e.g., semantic scene segmentation (e.g., [LMP01, PPI09, VT08]). Here, semantic segmentation describes the task of labeling each pixel of an image with a semantic category (e.g., "sky", "car", "street"). Closely related to Bayesian surprise (see Sect. 3.3.1), Klein et al. [KF11] use the Kullback-Leibler Divergence of the center and surround image patch histograms to calculate the saliency. Cheng et al. [CZM+11] use segmentation to define a regional contrast-based method, which simultaneously evaluates global contrast differences and spatial coherence. In general, we can differentiate between algorithms that rely on segmentation-based (e.g., [CZM+11, ADF10]) and pixel-based contrast measures (e.g., [AHES09, AS10, KF11]).

It has been observed in several eye tracking studies that human gaze fixation locations in natural scenes are biased toward the center of images and videos (see, e.g.,

[Bus35, Tat07, PN03]). One possible bottom-up cause of the bias is intrinsic bottom-up visual saliency as predicted by computational saliency models. One possible top-down cause of the center bias is known as photographer bias (see, e.g., [RZ99, PN03, Tat07]), which describes the natural tendency of photographers to place objects of interest in the center of their composition. In fact, what the photographer considers interesting may also be highly perceptually, bottom-up salient. Additionally, the photographer bias may lead to a viewing strategy bias [PLN02], which means that viewers may orient their attention more often toward the center of the scene, because they expect salient or interesting objects to be placed there. Thus, since in natural images and videos the distribution of objects of interest and thus saliency is usually biased toward the center, it is often unclear how much the saliency actually contributes in guiding attention. It is possible that people look at the center for reasons other than saliency, but their gaze happens to fall on salient locations. Therefore, this center bias may result in overestimating the influence of saliency computed by the model and contaminate the evaluation of how visual saliency may guide orienting behavior. Recently, Tseng et al. [TCC+09] were able to demonstrate quantitatively that center bias is correlated strongly with photographer bias and is influenced by viewing strategy at scene onset. Furthermore, e.g., they were able to show that motor bias had almost no effect. Here, motor bias refers to a preference of short saccades over long saccades [Tat07, TCC+09]. This can affect the distribution of fixated image locations in eye tracking experiments, because in most free viewing experiments the participants are asked to start viewing from a central image location [TCC+09] (e.g., for purpose of calibration or consistency).

Interestingly, although it is now a well-studied aspect of eye tracking experiments to such an extent that it has become an integral part of evaluation measures (see Sect. 3.2.1.6), the photographer bias has neither been thoroughly studied nor well modeled in the field of salient object detection. Most importantly, in Jiang et al.'s work [JWY+11] one of the criteria that characterize a salient object is that "it is most probably placed near the center of the image", which is justified with the "rule of thirds" (see, e.g., [Pet03]). Most recently, Borji et al. [BSI12] evaluated several salient object detection models and also performed tests with an additive Gaussian center bias and conclude that the resulting "change in accuracy is not significant and does not alter model rankings". But, this study neglected the possibility that well-performing models already have integrated, implicit biases.

Contributions

We provide an empirical justification why a Gaussian center bias is in fact beneficial for salient object detection in web images. Then, we show that implicit, undocumented biases are at least partially responsible for the performance of state-of-the-art algorithms and adapt the segmentation-based method by Cheng et al. [CZM+11] to remove its implicit center bias. This way, we achieve four goals: First, we could invalidate the statement that salient object detection is unaffected by a photographer or otherwise incurred center bias (see [BSI12]). Second, we could quantify the influence that an integrated center bias can have on salient object detection models. Third, we could improve the state-of-the-art in salient object detection on web images through the integration of an explicit, well-modeled center bias. Fourth, we derived the currently best performing unbiased algorithm. The latter aspect is especially interesting for many applications domains in which the image data is not biased by a photographer (e.g., autonomous robots and cars, or surveillance).

Furthermore, with respect to our work on target object detection in the presence of top-down guidance, our task is substantially different to all prior art on salient object detection: First, we try to integrate top-down information such as pointing gestures, gaze, and language. Second, we do not limit ourselves to web images and thus the image data does not necessarily have a photographer bias. Third, our target objects are substantially smaller compared to typical salient objects in the most important datasets (see [LSZ+07, AHES09]).

4.1.2.2 Visual Search

It has been shown that knowledge about the target object influences the saliency to speed-up the visual search (see [STET01, WHK+04]). However, not every piece of knowledge can influence the perceptual saliency. Instead, only specific information that refers to preattentive features allows such guidance [WHK+04]. For example, knowing the specific object or at least its color reduces the search slope, whereas categorical (e.g., "animal" or "post card") information typically does not provide top-down guidance (see [WHK+04]). Accordingly, in recent years, various computational saliency models have been developed that are able to integrate top-down knowledge in order to guide the attention in goal-directed search (e.g., [TCW+95, IK01b, FBR05, Fri06, NI07, WAD09, Wel11]). However, the number of saliency models that have been designed for goal-directed search is small compared to the vast amount of bottom-up saliency models (see Sects. 2.1.1 and 3.1.1), which might be symptomatic for the fact that there does not exist any established dataset to evaluate top-down visual search algorithms.

Most importantly, Navalpakkam and Itti [NI07] introduced a saliency model that allows to predict the visual search pattern given knowledge about the visual appearance of the target and/or distractors. In principle, this can be seen as an implementation of Wolfe et al.'s guided search model (GSM) [WCF89, Wol94], see Sect. 2.1.1. Navalpakkam and Itti use the knowledge about the target's appearance to maximize the expected signal-to-noise ratio (SNR), i.e. target-to-distractor ratio, of the saliency combination across and within feature dimensions. For this purpose, every feature dimension (e.g., orientation or intensity) is additionally subdivided by neurons with broadly overlapping Gaussian tuning curves to model varying neuron sensitivities to different value ranges within each feature dimension. Then, for each neuron's response the center-surround contrast is calculated to form each neuron's feature map. This way, it is possible to assign a higher or lower weight to salient aspects within value bands of each feature dimension; for example, to assign a higher importance to the response of neurons that encode very bright areas or roughly 45° edges.

Contributions

In contrast to prior art, we do not just focus on isolated non-verbal or verbal aspects. Instead, we integrate all available knowledge provided by different but complementing modalities in a computational model. Therefore, we use CRFs to integrate bottom-up saliency models, salient object detection methods, spatial corridors of attention given by gaze and pointing gestures, and—if available—spoken object descriptions. We also show that for our task the CRFs are able to significantly outperform neuron-based approaches such as Navalpakkam and Itti's model [NI07], which was adapted by Schauerte and Fink to use spectral saliency to calculate each neuron's feature map [SF10a]. We would like to note that Navalpakkam and Itti's neuron-based approach, albeit its age, is still the most established model in the field.

This way, we are often able to select the correct target object, even in complex situations. It is important to note that even for isolated aspects our latest datasets (i.e., ReferAT and Gaze@Flickr) are substantially more complex than what has been used by other research groups (compare, e.g., [Fri06, BK11] for visual search and [SKI+07, NS07] for pointed-at objects). An interesting aspect of our approach is that it is able to accurately segment the target object in most "simple" situations, although it can only predict salient regions of interest in complex situations.

4.2 Debiased Salient Object Detection

As we addressed in the introduction of this chapter, we are interested in two different data domains, see Fig. 4.2: First, human-robot interaction, in which the people try to direct a robot's attention. Second, web images, in which a photographer tries to highlight and direct our attention to certain aspects of the scene. Here, the photographer might (e.g., see the Gaze@Flickr dataset, Sect. 4.4.2) or might not (e.g., see the MSRA dataset, Sect. 4.2.1) use persons and visible non-verbal cues to guide the attention. Naturally, the images from the two domains follow different biases such as, for example, that the target objects in photographs are often substantially larger compared to the target objects in our human-robot interaction scenes.

(a)   **(b)**   **(c)**   **(d)**

Fig. 4.2 Example images of the **a** MSRA, **b** Gaze@Flickr, **c** PointAT, and **d** ReferAT datasets to illustrate the domain specific differences

Web images and photographs in general form a domain that is substantially different from images that are not directly composed by humans (e.g., surveillance footage, robot and unmanned aerial vehicle camera images, etc.). This is due to the fact that photographers follow image composition rules such as, for example, the rule of thirds (see, e.g., [Pet03]), which leads to very specific biases. We have already seen that such image composition biases have an important influence on saliency models, because one of the reasons why the AUC evaluation measure (Sect. 3.2.1.6) is favored by many researchers is that it compensate for such biases; most importantly, the center-bias that is commonly found in eye tracking datasets is linked to the photographer bias [TCC+09].

Nowadays, most work on salient object detection focuses on web images and MSRA is the dominant dataset in that research area. As a consequence, this means that we can expect that the algorithms are (over-)adapted to that specific domain. However, current state-of-the-art salient object detection algorithms are very powerful and, consequently, we would like to build on that work and derive an unbiased salient object detection algorithm that can help us as a feature for other application domains such as, for example, surveillance footage or robotics.

4.2.1 The MSRA Dataset

The most important salient object detection dataset is the MSRA dataset that has been created by Achanta et al. and Liu et al. [AHES09, LSZ+07], see Fig. 4.3. The MSRA dataset is based on the salient object dataset by Liu et al. [LSZ+07] and consists of a subset of 1000 image for which Achanta et al. provide annotated segmentation masks [AHES09], see Fig. 4.3. The images in Liu et al.'s dataset have been collected from a variety of sources, mostly from image forums and image search engines. Liu et al. collected more than 60,000 images and subsequently selected an image subset in which all images contain a salient object or a distinctive foreground object [LSZ+07]. Then, 9 users marked the salient objects using (rough) bounding boxes and the salient objects in the image database have been defined based on the "majority agreement". However, as a consequence of the selection process, the dataset does not include images without distinct salient objects and is potentially biased by the human selectors. This is an important aspect to consider when trying to generalize the results reported on Achanta et al.'s and Liu et al.'s dataset to other datasets or application areas.

Dataset Properties

The 1000 images contain 1265 annotated target object regions. On average a target object region occupies 18.25 % of the image area. Furthermore, as we will show in the following, the object locations in the dataset are strongly biased toward the center of the image.

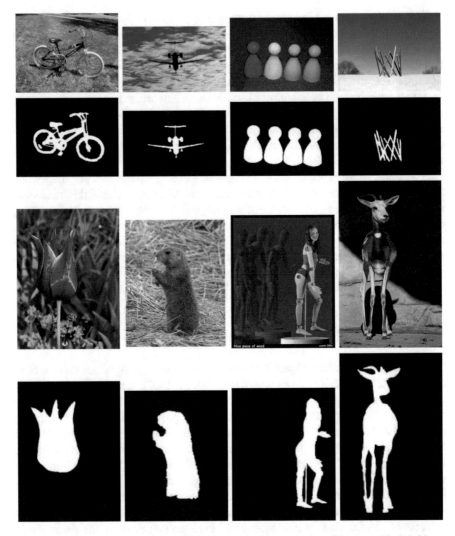

Fig. 4.3 Example images from the MSRA dataset. Depicted are selected images with their binary target segmentation masks

4.2.2 MSRA's Photographer Bias

To investigate the spatial distribution of salient objects in photographs, we use the segmentation masks by Achanta et al. [AHES09, AS10]. More specifically, we use the segmentation masks to determine the centroids of all salient objects in the dataset and analyze the centroids' spatial distribution.

4.2.2.1   Salient Object Distribution Model

The Center

Our model is based on a polar coordinate system that has its pole at the image center. Since the images in Achanta's dataset have varying widths and heights, we use in the following normalized Cartesian image coordinates in the range $[0, 1] \times [0, 1]$. The mean salient object centroid location is $[0.5021, 0.5024]^T$ and the corresponding covariance matrix is $[0.0223, -0.0008; -0.0008, 0.0214]$. Thus, we can motivate the use of a polar coordinate system that has its pole at $[0.5, 0.5]^T$ to represent all locations relative to the expected distribution's mode.

The Angles are Distributed Uniformly

Our first model hypothesis is that the centroids' angles in the specified polar coordinate system are uniformly distributed in $[-\pi, \pi]$.

   To investigate the hypothesis, we use a quantile-quantile (Q-Q) plot as a graphical method to compare probability distributions (see [NIS12]). In Q-Q plots the quantiles of the samples of two distributions are plotted against each other. Thus, the more similar the two compared distributions are, the better the points in the Q-Q plot will approximate the line $f(x) = x$. We calculate the Q-Q plot of the salient object location angles in our polar coordinate system versus uniformly drawn samples in $[-\pi, \pi]$, see Fig. 4.4a. The apparent linearity of the plotted Q-Q points supports the hypothesis that the angles are distributed uniformly.

The Radii Follow a Half-Gaussian Distribution

Our second model hypothesis is that the radii of the salient object locations follow a half-Gaussian distribution. We have to consider a truncated distribution in the interval $[0, \infty]$, because the radius—as a length—is by definition positive. If we consider the image borders, we could assume a two-sided truncated distribution, but we have three

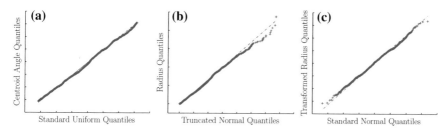

Fig. 4.4   Quantile-quantile (Q-Q) plots of the angles versus a uniform distribution (**a**), radii versus a half-Gaussian distribution (**b**), transformed radii (see Sect. 4.2.2.2) versus a normal distribution (**c**)

reasons to work with a one-sided model: The variance of the radii seems sufficiently small, the "true" centroid of the salient object may be outside the image borders (i.e., parts of the salient object can be truncated by the image borders), and it facilitates the use of standard statistical tests (see Sect. 4.2.2.2).

We use a Q-Q plot against a half-Gaussian distribution to graphically assess the hypothesis, see Fig. 4.4b. The linearity of the points suggests that the radii are distributed according to a half-Gaussian distribution. The visible outliers in the upper-right are caused by less than 30 centroids that are highly likely to be disturbed by the image borders. Please be aware of the fact that it is not necessary to know the half-Gaussian (or standard Gaussian) distribution's model parameters when working with Q-Q plots (see [NIS12]).

4.2.2.2 Empirical Hypothesis Analysis

We can quantify the observed linearity in the Q-Q plots, see Fig. 4.4, to analyze the correlation between the model distribution and the data samples using probability plot correlation coefficient (PPCC) [NIS12]. The PPCC is the correlation coefficient between the paired quantiles and measures the agreement of the fitted distribution with the observed data (i.e., goodness-of-fit). The closer the correlation coefficient is to one, the higher the positive correlation and the more likely the distributions are shifted and/or scaled versions of each other. By comparing against critical values of the PPCC (see [VK89, NIS12]), we can use the PPCC as a statistical test that is able to reject the hypothesis that the observed samples come from identical distributions. This is closely related to the Shapiro-Wilk test [SW65]. Furthermore, we can use the correlation to test the hypothesis of no correlation by transforming the correlation to create a t-statistic.

Although often data analysts prefer to use graphical methods such as Q-Q plots to assess the feasibility of a model, formal statistical hypothesis tests remain the most important method to disprove hypotheses. The goal of statistical tests is to determine if the (null) hypothesis can be rejected. Consequently, statistical tests either reject (prove false) or fail to reject (fail to prove false) a null hypothesis. But, they can never prove it true (i.e., failing to reject a null hypothesis does not prove it true). However, we can disprove alternate hypotheses and, additionally, we can use a set of statistical tests that are based on different principles. If all tests fails, we have—at least—an indicator that the hypothesis is potentially true.

The Angles are Distributed Uniformly

The obvious linearity of the Q-Q plot, see Fig. 4.4a, is reflected by a PPCC of 0.9988,[2] which is substantially higher than the critical value of 0.8880 (see [VK89]) and

---

[2]Mean of several runs with $N = 1000$ uniform randomly selected samples.

thus the hypothesis of identical distributions can not be rejected. Furthermore, the hypothesis of no correlation is rejected at $\alpha = 0.05$ ($p = 0$).

We use Pearson's χ^2 test [Pea00] as a statistical hypothesis test against a uniform distribution. The test fails to reject the hypothesis at significance level $\alpha = 0.05$ ($p = 0.2498$). Considering the circular type of data, we use Rayleigh's and Rao's tests for circular uniformity and both tests fail to reject the hypothesis at $\alpha = 0.05$ ($p = 0.5525$ and $p > 0.5$, respectively; see [Bat81]). On the other hand, for example, we can reject the alternative hypotheses of a normal or exponential distribution using the Lilliefors test [Lil67] ($p = 0$ for both distributions[3]).

The Radii Follow a Half-Gaussian Distribution

In order to use standard statistical hypothesis tests, we transform the polar coordinates in such a way that they represent the same point with a combination of positive angles in $[0, \pi]$ and radii in $[-\infty, \infty]$. According to our hypothesis, the distribution of the transformed radii should follow a normal distribution with its mode and mean at 0, see Fig. 4.4c.

The correlation that is visible in the Q-Q plot, see Fig. 4.4b, c, is reflected by a PPCC of 0.9987, which is above the critical value of 0.9984 (see [NIS12]). The hypothesis of no correlation is rejected at $\alpha = 0.05$ ($p = 0$).

Again we disprove exemplary alternate hypotheses: The uniform distribution is rejected by the test against the critical value of the PPCC as well as by Pearson's χ^2 test at $\alpha = 0.05$ ($p = 0$). The exponential distribution is rejected by Lilliefors test at $\alpha = 0.05$ ($p = 0$). We perform the Jarque-Bera, Lilliefors, Spiegelhalter's, and Shapiro-Wilk test (see [BJ80, Lil67, Spi83, SW65]) to test our null hypothesis that the radii have been sampled from a normal distribution (unknown mean and variance). Subsequently, we use a T-test to test our hypothesis that the mean of the radius distribution is 0. The Jarque-Bera, Lilliefors, Spiegelhalter's, and Shapiro-Wilks tests fail to reject the hypothesis at significance level $\alpha = 0.05$ ($p = 0.8746$, $p = 0.2069$, $p = 0.2238$, and $p = 0.1022$, respectively). Furthermore, it is likely that the mode of the (transformed) radius is 0, because the corresponding T-test fails to reject the hypothesis at significance level $\alpha = 0.05$ with $p = 0.9635$.

4.2.3   Salient Object Detection

4.2.3.1   Algorithm

We adapt the region contrast model by Cheng et al. [CZM+11]. Cheng et al.'s model is particularly interesting, because it already provides state-of-the-art performance,

---

[3]We report $p = 0$, if the tabulated values are 0 or the Monte Carlo approximation returns 0 or ϵ (double-precision).

Fig. 4.5 An example of the influence of the center bias on segmentation-based salient object detection. *Left-to-right* example image from the MSRA dataset, region contrast without and with center bias (RC'10 and RC'10+CB, resp.), and locally debiased region contrast without and with center bias (LDRC and LDRC+CB, resp.)

which is partially caused by an implicit center bias. Thus, we can observe how the model behaves if we remove the implicit center bias, which was neither motivated nor explained by the authors, and add an explicit Gaussian center bias (Fig. 4.5). We extend the spatially weighted region contrast saliency equation S_{RC} (see Eq. 7 in [CZM+11]) and integrate an explicit, linearly weighted center bias:

$$S_{RC+CB}(r_k) = w_B S_{RC}(r_k) + w_C\, g(C(r_k); \sigma_x, \sigma_y) \quad \text{with} \tag{4.1}$$

$$S_{RC}(r_k) = \sum_{r_k \neq r_i} \hat{D}_s(r_k; r_i) w(r_i) D_r(r_k; r_i) \quad \text{and} \tag{4.2}$$

$$\hat{D}_s(r_k; r_i) = \exp(-D_s(r_k; r_i)/\sigma_s^2) . \tag{4.3}$$

Here, we use a convex combination[4] to control the strength of the influence of the center bias, i.e. $w_B + w_C = 1$ ($w_B, w_C \in \mathbf{R}_0^+$). $\hat{D}_s(r_k; r_i)$ is the spatial distance between regions r_k and r_i, where σ_s controls the spatial weighting. Smaller values of σ_s influence the spatial weighting in such a way that the contrast to regions that are farther away contributes less to the saliency of the current region. The spatial distance between two regions is defined as the Euclidean distance between the centroids of the respective regions using pixel coordinates that are normalized to the range $[0, 1] \times [0, 1]$. Furthermore, $w(r_i)$ is the weight of region r_i and $D_r(\cdot; \cdot)$ is the color distance metric between the two regions (see [CZM+11] for more details). Here, the number of pixels in r_i is used as $w(r_i) = |r_i|$ to emphasize color contrast to bigger regions. $C(r_k)$ denotes the centroid of region r_k and g is defined as follows

$$g(x, y; \sigma_x, \sigma_y) = \frac{1}{\sqrt{2\pi}\sigma_x} \exp\left\{-\frac{1}{2}\frac{x^2}{\sigma_x^2}\right\} * \frac{1}{\sqrt{2\pi}\sigma_y} \exp\left\{-\frac{1}{2}\frac{y^2}{\sigma_y^2}\right\}. \tag{4.4}$$

Interestingly, the unnormalized Gaussian weighted Euclidean distance used by Cheng et al. [CZM+11] causes an implicit Gaussian-like center bias, see Fig. 4.6, because it favors regions whose distances to the other neighbors are smaller. Unfortunately, this has not been motivated, discussed, or evaluated by Cheng et al. To remove

---

[4]We have considered different combination methods and provide a quantitative evaluation of different combination types in Appendix D. In short, the convex combination achieves the best overall performance of the evaluated combination methods.

(a)   **(b)**

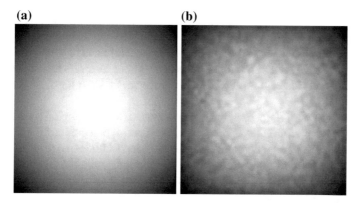

Fig. 4.6 Illustration of the implicit center bias in the method by Cheng et al. [CZM+11]. **a** Each pixel shows the distance weight sum, i.e. $\sum_{r_i} \hat{D}_s(r_k; r_i)$, to all other pixels in a regular grid. **b** The average weight sum depending on the centroid location calculated on the MSRA dataset

this implicit bias, we introduce a normalized, i.e. locally debiased, distance function $\check{D}_s(r_k; r_i)$ that still weights close-by regions higher than further away regions, but does not lead to an implicit center bias

$$\check{D}_s(r_k; r_i) = \frac{\hat{D}_s(r_k; r_i)}{\sum_{r_j} \hat{D}_s(r_k; r_j)}, \tag{4.5}$$

$$\text{i.e.} \quad \forall r_k : \sum_{r_i} \check{D}_s(r_k; r_i) = 1, \tag{4.6}$$

and define

$$S_{\text{LDRC}}(r_k) = \sum_{r_k \neq r_i} \check{D}_s(r_k; r_i) w(r_i) D_r(r_k; r_i) \quad \text{and} \tag{4.7}$$

$$S_{\text{LDRC+CB}}(r_k) = w_{\text{B}} S_{\text{LDRC}}(r_k) + w_{\text{C}} \, g(C(r_k); \sigma_x, \sigma_y). \tag{4.8}$$

4.2.3.2 Evaluation

Baseline Algorithms

To compare our results, we use a set of state-of-the-art salient object detection algorithms: Achanta et al.'s frequency-tuned model (AC'09; [AHES09]), Achanta et al.'s maximum symmetric surround saliency model (MSSS'10; [AS10]), Klein et al.'s information-theoretic saliency model (BITS'11; [KF11]), and Cheng et al.'s region contrast model (RC'10; [CZM+11]) that uses Felzenszwalb's image segmentation method [FH04]. The latter is the original algorithm that we adapted in Sect. 4.2.3.1.

Additionally, we include the results of four algorithms that have not been designed for salient object detection: Itti and Koch's model [IKN98] as implemented by Itti's iLab Neuromorphic Vision Toolkit (iNVT'98), Harel et al.'s graph-based visual saliency (GBVS'07; [HKP07]), Goferman et al.'s context-aware saliency (CAS'12; [GZMT12]), and Guo et al.'s pure Fourier transform (PFT'07; [GMZ08, GZ10]; see Sect. 3.2.1.1).

To investigate the influence of the implicit center bias in the region contrast model, we calculate the performance of the locally debiased region contrast (LDRC) model without and with our explicit center bias (LDRC and LDRC+CB, respectively). For comparison, we also evaluate the region contrast model with the additional explicit center bias (RC'10+CB). As additional baseline, we provide the results for simple segment-based and pixel-based—i.e., using Eq. 4.4 for each pixel with respect to the image center distance and constant variance—center bias models, i.e. $w_C = 1$ (CB$_S$ and CB$_P$, respectively).

Measures

We can use the binary segmentation masks for saliency evaluation by treating the saliency maps as binary classifiers. At a specific threshold t we regard all pixels that have a saliency value above the thresholds as positives and all pixels with values below the thresholds as negatives. By sweeping over all thresholds $\min(S) \leq t \leq \max(S)$, we can evaluate the performance using common binary classifier evaluation measures.

We use four evaluation measures to quantify the performance of the evaluated algorithms. We calculate the area under the curve (AUC) of the ROC curve ($\int$ROC). Complementary to the $\int$ROC, we calculate the maximum F_1 and $F_{\sqrt{0.3}}$ scores (see [AHES09] and Sect. 3.3.2.1). F_β with $\beta = \sqrt{0.3}$ has been proposed by Achanta et al. to weight precision higher than recall for salient object detection [AHES09]. Additionally, we calculate the PHR, see Sect. 4.3.1.5, which measures how often the pixel with the maximum saliency belongs to a part of the target object.

Results

The performance of RC'10 drops substantially if we remove the implicit center bias as is done by LDRC, see Table 4.1. However, if we add our explicit center bias model to the unbiased model, the performance is substantially increased with respect to all evaluation measures. Furthermore, with the exception of pixel hit rate (PHR), the performance of LDRC+CB and RC'10+CB is nearly identical with a slight advantage for LDRC+CB. This indicates that we did not lose important information by debiasing the distance metric (LDRC+CB vs. RC'10+CB) and that the explicit Gaussian center bias model is advantageous compared to the implicit weight bias (LDRC+CB and RC'10+CB versus RC'10).

Table 4.1 The maximum F_1 score, maximum F_β score, ROC AUC ($\int$ROC), and PHR of the evaluated algorithms (sorted by descending F_β)

Method	F_β	F_1	$\int$ROC	PHR
LDRC+CB	0.8183	0.8034	0.9624	0.9240
RC'10+CB	0.8120	0.7973	0.9620	0.9340
RC'10	0.7993	0.7855	0.9568	0.9140
LDRC	0.7675	0.7574	0.9430	0.8680
BITS'11	0.7582	0.7342	0.9316	0.7540
MSSS'10	0.7337	0.7165	0.9270	0.8420
GBVS'07	0.6242	0.6403	0.9088	0.8480
PFT'07	0.6009	0.5995	0.8392	0.7100
CB_S	0.5764	0.5793	0.8623	0.6980
CAS'12	0.5615	0.5857	0.8741	0.6920
CB_P	0.5452	0.5604	0.8673	0.7120
iNVT'98	0.4012	0.3383	0.5768	0.6870

Most interestingly, LDRC is the best model without center bias, which makes it interesting for applications in which the image data can not be expected to have a photographer's center bias (e.g., image data of surveillance cameras or autonomous robots).

4.2.4 Debiased Salient Object Detection and Pointing

Although it is interesting that we were able to improve the state-of-the-art in salient object detection by analyzing and explicitly modeling task specific biases, see Sect. 4.2.3.2, our initial motivation has been to improve the ability of salient object detection algorithms to perform well in other domains such as, e.g., our PointAT and ReferAT pointing datasets. We can see in Table 4.2 that our debiased LDRC algorithm leads to substantial performance improvements in predicting the right target objects in three conditions: First, without any directional information from the pointing gestures, see Table 4.2 "none". Second, in combination with a heuristic model—the heuristic method will be described in the subsequent Sect. 4.3—to integrate automatically detected pointing gestures, see Table 4.2 "automatic". Third, in combination with the heuristic method and manually annotated pointing information, see Table 4.2 "annotated".

However, we will see in the following section that LDRC itself is not the best predictor for this task and its performance is surpassed by our spectral saliency models (see Sect. 3.2). This is most likely caused by the small target object sizes

Table 4.2 Target object prediction results of RC'10 and its debiased counterpart LDRC on the PointAT and ReferAT datasets

Algorithm	PHR (%)	FHR (%)	$\int$FHR	NTOS
(a) PointAT				
	None			
RC'10	0.94	1.88	0.029	0.659
LDRC	2.83	4.24	0.045	0.792
	Automatic			
RC'10	53.77	67.92	0.817	5.901
LDRC	58.49	74.53	0.828	5.831
	Annotated			
RC'10	61.79	78.30	0.843	6.596
LDRC	60.85	80.66	0.852	6.431
(b) ReferAT				
	None			
RC'10	0.00	1.65	0.025	0.226
LDRC	0.00	2.47	0.026	0.427
	Automatic			
RC'10	2.48	19.42	0.421	3.190
LDRC	4.96	28.51	0.551	3.333
	Annotated			
RC'10	4.13	23.97	0.502	3.778
LDRC	7.02	30.99	0.621	3.916

A description of the evaluation measures can be found in Sect. 4.2.3.2

that are atypical for common salient object detection tasks. Nonetheless, LDRC will turn out to be a useful feature that helps us to highlight the right target object area. Furthermore, being able to control the center bias allows us to seamlessly disable the center bias for PointAT and ReferAT while adding a center bias when processing Gaze@Flickr in Sect. 4.4.

4.3 Focusing Computational Attention in Human-Robot Interaction

As briefly mentioned in the introduction to this chapter, verbal and non-verbal signals that guide our attention are an essential aspect of natural interaction and help to establish a joint focus of attention (e.g., [Ban04, BI00, GRK07, Piw07, Roy05, SC09]). In other words, the ability to generate and respond to attention directing signals allows to establish a common point of reference or conversational domain with an interaction partner, which is fundamental for "learning, language, and sophisticated social competencies" [MN07b].

When talking about the focus of attention (FoA) in interaction, we have to distinguish between the FoA within the conversation domain (i.e., what people are talking about), and the perceptual focus of attention (e.g., where people are looking at). In many situations, the conversational FoA and the perceptual FoA can and will be distinct. However, when persons are referring to specific objects within a shared spatial environment, multimodal—here, non-verbal and verbal—references are an important part of natural communication to direct the perceptional FoA toward the "referent", i.e. the referred-to object, and achieve a shared conversational FoA. Accordingly, we have to distinguish between the saliency of objects in the context of the conversation domain at some point during the interaction and the inherent, perceptual saliency of objects present in the scene (see [BC98]). Although the conversational domain is most important when identifying the referent—especially when considering object relations—, the perceptual saliency can influence the generation and interpretation of multimodal referring acts to such extend that in some situations "listeners [...] identify objects on the basis of ambiguous references by choosing the object that was perceptually most salient" [BC98, CSB83].

In this section, we focus on a situation in which an interacting person uses non-verbal (i.e., pointing gestures) and verbal (i.e., specific object descriptions) signals to direct the attention of an interaction partner toward a target object. Consequently, our task is to highlight the intended target object. One of the challenges in such a situation is that we may know nothing about the target object's appearance. In fact, it might be the actual goal of the multimodal reference to teach something about the referent; for example, imagine a pointing gesture that is accompanied by an utterances like "Look at that! I bet you have never seen Razzmatazz[5] before!".

Interestingly, exactly in situations in which we know nothing about the target object's appearance, we can use visual saliency models (see Sects. 3.2 and 4.2) to determine image regions that are highly likely to render potential objects of interest. And, as we have mentioned before, the actual multimodal reference itself might even be influenced by the perceptual saliency. Furthermore, some target information (e.g., information about the target's color) is known to subconsciously influence our visual attention system and accordingly we can try to integrate such information as well, if it is available. For both situations, i.e. pointing gestures in the absence (Sect. 4.3.1) or presence (Sect. 4.3.2) of verbal object descriptions, we present two approaches: First, purpose-built heuristic and neuron-based models that were specifically proposed for this purpose by Schauerte and Fink. Second, we use machine learning with conditional random fields, which has two advantages: This approach leads to a better, more robust predictive performance and it is simpler to integrate additional features and target information.

---

[5]Razzmatazz is a color name and describes a shade of rose or crimson.

4.3.1 Pointing Gestures

Like gaze, pointing gestures direct the attention into the visual periphery, which is indicated by the pointing direction (see Sect. 4.1). The pointing gesture's directional information is defined by the origin o—usually the hand or finger—and an estimation of the direction d. The referent, i.e. the object that is being pointed at, can then be expected to be located in the corridor of attention alongside the direction. However, the accuracy of the estimated pointing direction depends on multiple factors: the inherent accuracy of the performed gesture (see [BO06, BI00, KLP+06]), the method to infer the pointing direction (see [NS07]), and the automatic pointing gesture detection itself.

4.3.1.1 Pointing Gesture Detection

In the following, we briefly describe how we detect pointing gestures, calculate the indicated pointing direction, assess the detected pointing gesture's inaccuracy, and model how likely the pointing gesture directs the attention to specific image locations. As has been done by Nickel and Stiefelhagen [NS07], we use the line-of-sight model to calculate the indicated pointing direction. In this model, the pointing direction is equivalent to the line-of-sight defined by the position of the eyes h and the pointing hand o, and accounts for "the fact that [in many situations; A/N] people point by aligning the tip of their pointing finger with their dominant eye" [BO06]. To recognize pointing gestures, we adapted Richarz et al.'s approach [RPF08].[6] Without going into any detail, we replaced the face detector with a head shoulder detector based on histogram of oriented gradients (HOG) features to improve the robustness. We detect the occurrence of a pointing gesture by trying to detect the inherent holding phase of the pointing hand. Therefore, we group the origin o_t and direction d_t hypotheses over time t and select sufficiently large temporal clusters to detect pointing occurrences.

We consider three sources of pointing inaccuracy: Due to image noise and algorithmic discontinuities the detected head-shoulder rectangles exhibit a position and scaling jitter. In order to model the uncertainty caused by estimating the eye position from that detection rectangle r_t (at time t), we use a Normal distribution around the detection center $\bar{r}_t$ to model the uncertainty of the estimated eye position

$$p_e(x|r_t) = \mathcal{N}(\bar{r}_t, \sigma_e^2). \tag{4.9}$$

σ_e is chosen so that one quarter of $\bar{s}$ is covered by $2\sigma_e$, i.e.

$$\sigma_e = \bar{s}/8, \tag{4.10}$$

---

[6]Please note that the approaches presented in this section are independent of the actual pointing gesture recognition method—e.g., we are currently working toward using Microsoft Kinect—as long as it is possible to derive an angular (in-)accuracy measure.

where $\bar{s}$ is the mean of the detection rectangle's size over the last image frames. Furthermore, we consider the variation in size of the head-shoulder detection rectangle, and the uncertainty of the estimated pointing direction d, which is caused by shifts in the head and hand detection centers. We treat them as independent Gaussian noise components and estimate their variances σ_s^2 and σ_d^2. As σ_e^2 and σ_s^2 are variances over positions, we approximately transfer them into an angular form by normalizing with the length $r = \|d\|$

$$\tilde{\sigma}_e^2 = \frac{\sigma_e^2}{r^2} \quad \text{and} \quad \tilde{\sigma}_s^2 = \frac{\sigma_s^2}{r^2}, \tag{4.11}$$

respectively. This approximation has the additional benefit to reflect that the accuracy increases when the distance to the pointer decreases and the arm is outstretched.

Spatial Pointing Target Probability

Due to Eq. 4.11, the combined accuracy distribution has become a distribution over angles

$$p_G(x) = p(\alpha(x; o, d)|d, o) = \mathcal{N}(0, \sigma_c^2), \tag{4.12}$$

with $\alpha(x; o, d)$ being the angle between the vector from the pointing origin o to the image point x given the pointing direction d, and

$$\sigma_c^2 \approx \tilde{\sigma}_e^2/r^2 + \tilde{\sigma}_s^2/r^2 + \sigma_d^2. \tag{4.13}$$

This equation models the probability $p_G(x)$ that a point x in the image plane was referred-to by the pointing gesture given the current head-shoulder detection d and the pointing direction o, and thus defines our corridor of attention. To account for the findings by Kranstedt et al. [KLP+06], we enforce a lower bound of $3°$, i.e.

$$\hat{\sigma}_c = \max(3°, \sigma_c), \tag{4.14}$$

so that 99.7 %, which corresponds to 3σ, of the distribution's probability mass covers at least a corridor of $9°$.

4.3.1.2 Heuristic Integration

Given a detected pointing gesture and the spatial target probability $p_G(x)$, see Eq. 4.12, we can calculate a top-down feature map

$$S_t(a, b) = p_G(x) \tag{4.15}$$

with $x = (a, b)$. This, in effect, defines a blurred cone of Gaussian probabilities— which encode how likely it is that the referent is located at the pixel's

(a)   **(b)**   **(c)**   **(d)**

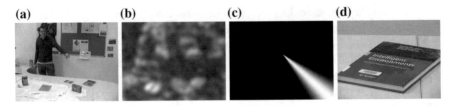

Fig. 4.7 Heuristic detection of the referent. **a** Input image, **b** bottom-up saliency, **c** probabilistic pointing cone, **d** attended object

position—emitted from the hand along the pointing direction in the image plane, see Fig. 4.7c. Due to its form, we refer to this map as being the probabilistic pointing cone (PPC).

We can now use visual saliency to act as a form of generalized object detector. For this purpose, we use QDCT to calculate the visual saliency map S_b based on the PCA decorrelated CIE Lab color space, see Sect. 3.2. The saliency map S_b is normalized to [0, 1] and each pixel's value is interpreted as being the probability that this pixel is part of the target object.

The final saliency map S is then obtained by calculating the joint probability

$$S = S_b \circ S_t \,, \tag{4.16}$$

where $\circ$ represents the Hadamard product. In other words, we highlight image regions alongside the pointing direction that are highly likely to contain a salient (proto-)object.

4.3.1.3 Conditional Random Field

Structure, Learning, and Prediction

In general, a CRF models the conditional probabilities of x (here, "does this pixel belong to a target object?"), given the observation y (i.e., features), i.e.

$$p(x|y) = \frac{1}{Z(y)} \prod_{c \in C} \psi(x_c, y) \prod_{i \in V} \psi(x_i, y), \tag{4.17}$$

where C is the set of cliques in the CRF's graph and i represent individual nodes. Here, ψ indicates that the value for a particular configuration x_c depends on the input y.

Naturally, our problem is a binary segmentation task, since the location depicted by a pixel can either belong to the target object or not, i.e. x_i can either be "target" or "background". We use a pairwise, 4-connected grid CRF structure. We linearly parametrize the CRF parameter vector Θ in unary node $u(y, i)$ (i.e., information at an image location) and edge features $v(y, i, j)$ (e.g., relating neighbored image locations). Here, it is important to consider that the cliques in a 4-connected, grid-structured graph are the sets of connected nodes, which are represented by the edges. Thus, we fit two matrices F and G such that

$$\Theta(x_i) = Fu(y, i) \tag{4.18}$$
$$\Theta(x_i, x_j) = Gv(y, i, j). \tag{4.19}$$

Here, y is the observed image and $\Theta(x_i)$ represents the parameter values for all values of x_i. Similarly, $\Theta(x_i, x_j)$ represents the parameter values for all x_i, x_j. Then, we can calculate

$$p(x; \Theta) = \exp\left[\sum_i \Theta(x_i) + \sum_j \Theta(x_i, x_j) - A(\Theta)\right], \tag{4.20}$$

where $A(\Theta)$ is the log-partition function that ensures normalization.

We use tree-reweighted belief propagation (TRW) to perform approximate marginal inference, see [WJ08]. TRW addresses the problem that it is computationally intractable to compute the log-partition function $A(\Theta)$ exactly and approximates $A(\Theta)$ with

$$\hat{A}(\Theta) = \max_{\mu \in \mathcal{L}} \Theta \cdot \mu + \hat{H}(\mu), \tag{4.21}$$

where $\hat{H}$ is TRW's entropy approximation [WJ08]. Here, $\mathcal{L}$ denotes the valid set of marginal vectors

$$\mathcal{L} = \{\mu : \sum_{x_{c\backslash i}} \mu(x_c) = \mu(x_i) \wedge \sum_{x_i} \mu(x_i) = 1\}, \tag{4.22}$$

where μ describes a mean vector, which equals a gradient of the log-partition function. Then, the approximate marginals $\hat{\mu}$ are the maximizing vector

$$\hat{\mu} = \arg\max_{\mu \in \mathcal{L}} \Theta \cdot \mu + \hat{H}(\mu). \tag{4.23}$$

This can be approached iteratively until convergence or a maximum number of updates [Dom13].

To train the CRF, we rely on the clique loss function, see [WJ08],

$$L(\Theta, x) = -\sum_c \log \hat{\mu}(x_c; \Theta). \tag{4.24}$$

Here, $\hat{\mu}$ indicates that the loss is implicitly defined with respect to marginal predictions—again, in our implementation these are determined by TRW—and not the true marginals. This loss can be interpreted as empirical risk minimization of the mean Kullback-Leibler divergence of the true clique marginals to the predicted ones.

Features

As unary image-based features, we include the following information at each CRF grid point: First, we include each pixel's normalized horizontal and vertical image position in the feature vector. Second, we directly use the pixel's intensity value after scaling the image to the CRF's grid size. Third, we include the scaled probabilistic pointing cone (PPC), see Sect. 4.3.1.1. Then, after scaling each saliency map to the appropriate grid size, we append QDCT image signature saliency maps based on the PCA decorrelated Lab color space (see Sect. 3.2.1) at three scales: $96 \times 64\,px$, $168 \times 128\,px$, and $256 \times 192\,px$. Optionally, we include the LDRC salient object prediction, see Sect. 4.2.

As CRF edge features, first, we use a constant of one that allows to model general neighborhood relations. Second, we use 10 thresholds to discretize the L^2 norm of the color difference and thus contrast of neighboring pixels. Then, we multiply the existing features by an indicator function for each edge type (i.e., vertical and horizontal), effectively doubling the number of features and encoding conjunctions of features and edge type. This way, we parametrize vertical and horizontal edges separately [Dom13].

4.3.1.4   The PointAT Dataset

To assess the ability of systems to identify arbitrary target objects in the presence of a pointing gesture, we collected a dataset that contains 220 images of 3 persons pointing at various objects [SRF10]. The dataset was recorded in two environments (an office space and a conference room) with a large set of objects of different category, shape, and texture. The dataset focuses on the object detection and recognition capabilities and was not supposed to be used to evaluate the performance of pointing gesture detection, which explains the limited number of pointing persons.

Hardware Setup

The dataset was recorded using a monocular Sony EVI-D70P pan-tilt-zoom camera. The camera provides images in approximately PAL resolution ($762 \times 568\,px$), and offers an optical zoom of up to $18\times$. Its wide horizontal opening angle is $48°$. To reflect a human or humanoid point of view, we mounted the camera on eye height of an averagely tall human [MFOF08].

Fig. 4.8 Examples of pointing gestures performed in the evaluation. Depicted are some exemplary images with the corresponding binary target object masks that we generate based on the annotated target object boundaries

Procedure

Each person performed several pointing sequences, with varying numbers and types of objects present in the scene. We neither restricted the body posture of the subjects in which pointing gestures had to be performed, nor did we define fixed positions for the objects and persons. The only restriction imposed was that the subjects were instructed to point with their arms extruded, so that pointing gestures would comply with the line-of-sight model employed. To evaluate the ability of the iterative shift of attention to focus the correct object in the presence of distractors, we occasionally arranged clusters of objects so that the object reference would be ambiguous. Accordingly, the dataset contains a wide variety of pointing references, see Fig. 4.8. Since we do not specifically evaluate the pointing detector (see [RPF08]), we discard cases with erroneous pointing gesture detections. Thus, in total, our evaluation set contains 220 object references.

For each object reference, we manually annotated the target object's boundaries.[7] Furthermore, we annotated the dominant eye, the pointing finger, and the resulting pointing direction. This makes it possible to assess the influence that the automatic pointing gesture recognition's quality has on the identification of the pointed-at object.

Properties

On average the target object occupies only 0.58 % of the image area. In other words, at random we would require roughly 200 trials to expect to select one pixel that is part of the target object. The average differences between the annotated and automatically

---

[7]Please note that we re-annotated the original dataset to calculate the results presented in the dissertation [Sch14] and this book, because the original annotations only consisted of bounding boxes. Accordingly, our results are not directly comparable to previously reported results [SRF10].

determined pointing origin and direction are 15.30 px and 2.95°, respectively. The former is mostly caused by the fact that the system detects the center of the hand, instead of the finger. The latter is due to the fact that the eye positions are estimated given the head-shoulder detection (see [RPF08]), and that the bias introduced by the dominant eye is unaccounted for (see [BO06]). In some cases, the ray that originates from the finger tip (i.e., the pointing origin) and follows the pointed direction does not intersect the target object's boundaries, i.e. it "misses" the object. The rate of how often the object's annotated boundary polygon is missed by the pointing ray is 5.66 % for the annotated pointing information and 19.34 % for the automatic detection.

4.3.1.5 Evaluation Measures

The fovea is responsible for detailed, sharp central vision. As such, it is essential for all tasks that require visual details such as, most importantly, many recognition tasks. However, the fovea itself comprises less than 1 % of the retinal area and can only perceive the central 2° of the visual field. In the following, we define that an object has been perceived or "focused", if it or a part of it has been projected onto the fovea. To make our evaluation independent of the recording equipment and image resolution— an important aspect for our evaluation on the Gaze@Flickr dataset (Sect. 4.4.2)—, we assume that a (hypothetical) human observer sits in front of a display on which the image is shown in full screen mode. This way, we can estimate the extent of the display's—and thus the image's—area that would be projected onto the model observer's fovea. Here, we assume a circular fovea area and, thus, a circular model FoA. We approximate the radius r_{FoA} of the FoA on the display as follows

$$ r_{\text{FoA}} = \tan(a_{\text{FoA}}/2) \frac{\sqrt{w^2 + h^2}}{D} d, \tag{4.25} $$

where a_{FoA} is the angle perceived by the fovea's visual field, d is the distance between the viewer's eyes and the screen, and D is the display's diameter. For example, this results in an FoA radius of 7.5 px for a fovea angle of 2°, a viewing distance of 65 cm—not untypical for office environments—, a 60.96 cm (24 in.) display diagonal, and an image resolution of 320 × 240 px. In the following, we define a fovea angle of 2°, a viewing distance of 65 cm, and a 24 in. display diagonal for all evaluations.

We derive two related evaluation measures: The pixel hit rate (PHR) measures how often the most salient pixel lies within the object boundaries (see [SS13a, SF10a, SRF10]). The focus of attention hit rate (FHR) measures how often the object is covered—and thus perceived—at least partially by the FoA (see [SF10a, SRF10]). To compute the FoA hit rate (FHR), we calculate whether the radial FoA and the annotated object's boundary polygon collide. Here, the assumption of an FoA area (i.e., FHR) instead of a simple FoA point (i.e., PHR) has an important benefit: Since saliency models tend to highlight edges, the most salient point is often related to the object's boundaries and as a consequence can be located just a bit outside of the actual object, very close to the boundary.

Additionally, we can calculate the FHR after shifting the focus of attention to the next most salient region in the image. To this end, we inhibit the location that has already been attended, i.e. we set the saliency of all pixels within the current FoA to zero. Let FHR^{+k} denote the FoA hit rate after k attentional shifts, i.e. how frequently the target is perceived within the first k shifts of attention. Then, we can integrate over the FHR^{+k} until a given $k \leq n$ (in the following, we set $n = 10$). We refer to this measure as $\int FHR$ and it has the advantage that it also reflects the cases in which the target object has not been found after n shifts.

Furthermore, similar to the normalized scanpath saliency (NSS) saliency measure (see Appendix C; [PIIK05, PLN02]), we calculate the mean saliency of the pixels that are part of the object area to compare the target area's saliency to the background saliency. This measure has a different purpose than PHR and FHR, because it does not directly evaluate the ability to focus the target object. Instead, it measures how strong the target object is highlighted against—i.e., separated from—the background. We call this measure normalized target object saliency (NTOS). For this purpose, the saliency map is normalized to have zero mean and unit standard deviation [PIIK05, PLN02], i.e. a NTOS of 1 means that the saliency in the target object area is one standard deviation above average. Consequently, an NTOS ≥ 1 indicates that saliency map has significantly higher saliency values at target object locations. An NTOS ≤ 0 means that the saliency does not predict a target object location better than picking random image locations.

4.3.1.6 Evaluation Results

Procedure

To train and evaluate our models, we use a leave-one-person-out training procedure (see Fig. 4.9). Furthermore, we mirror the samples along the vertical axis to double the available image data. The CRF is trained with a grid resolution of 381×284 and a 4-connected neighborhood.

Results

As can be seen in Table 4.3, CRFs provide a better predictive performance than the heuristic baseline method (see Sect. 4.3.1.2). Most interestingly, the CRF that we trained and tested with automatic gesture detections is able to outperform the heuristic method even if the latter relies on groundtruth pointing information. This shows that the CRF model is better capable to compensate for the inaccuracy of automatically detected pointing gestures. Accordingly, if we compare the performance of both methods with detected and annotated pointing information, we can see that the relative performance difference of the heuristic model is much higher than the performance difference for the CRF. Furthermore, we can see that LDRC in addition to our spectral features helps us to improve the results of the overall approach.

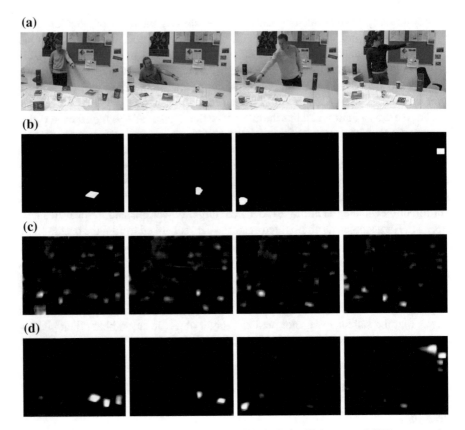

Fig. 4.9 Examples of pointing gestures performed in the PointAT dataset and CRF target region predictions. **a** Images, **b** masks, **c** without pointing information, **d** with groundtruth pointing information

To serve as a baseline, we asked human observers to guess the pointed-at object and they were able to estimate the correct object for about 87 % of the images [SRF10]. Accordingly, we can see that our model is able to come close to this baseline in terms of PHR. However, we can also see that the FHR is in fact higher than those 87 %. How can that be? Most importantly, in ambiguous situations in which two potential target objects stand close to each other, the predicted target object location tends to be between both objects or just on the point of a border of one object that is closest to the other object. Thus, the target object might not have been selected by the most salient pixel, but at least a part of it is visible in the assumed FoA.

Table 4.3 Target object detection performance on the PointAT dataset

Algorithm	PHR (%)	FHR (%)	∫FHR	NTOS
	No pointing			
Heuristic	7.54	10.84	0.238	2.094
CRF, no LDRC	9.43	16.04	0.257	2.110
CRF, w/LDRC	10.38	17.92	0.246	2.045
	Pointing detected			
Heuristic	59.91	82.08	0.838	7.425
CRF, no LDRC	80.66	92.45	0.879	9.727
CRF, w/LDRC	81.60	92.45	0.874	9.734
	Pointing annotated			
Heuristic	77.83	88.68	0.859	8.143
CRF, no LDRC	84.91	91.9	0.877	10.189
CRF, w/LDRC	85.38	92.92	0.878	10.083

A description of the evaluation measures can be found in Sect. 4.3.1.5

4.3.2 Language

Although pointing gestures are an important aspect of natural communication, there exist ambiguous situations in which it is not possible to identify the correct target object based on the pointing gesture alone. Instead, we commonly require additional knowledge or a shared context with our interaction partner to make the right assumptions about the referent. Although the combined use of gestures and language depends on the referring persons [Piw07], linguistic and gestural references can be seen to form composite signals, i.e. as one signal becomes more ambiguous the speaker will less rely on it and compensate with the other (see, e.g., [Ban04, BC98, GRK07, KLP+06, LB05, Piw07, SKI+07]). Accordingly, we now want to go a step further and not just react to pointing gestures, but also integrate spoken information about a target object's visual appearance. For this purpose, we have to make it possible to use the available information to guide the attention and highlight the intended target object.

4.3.2.1 Language Processing

To automatically determine spoken target information, we have to process and analyze the spoken utterance to determine references to object attributes or known objects. Before we proceed, please let us note that our intention is not to implement perfect language processing capabilities. Instead, we only want to assess how automatic—and thus sometimes faulty—extraction of target object information can influence our model's performance.

Target Information

Language often provides the discriminating context to identify the referent amidst other potential target objects. Most importantly, it is used to specify objects (e.g., "my Ardbeg whisky package"), classes (e.g., "whisky package"), visually deducible attributes (e.g., "red", or "big"), and/or relations (e.g., "the cup on that table"). When directly referring to an object, this information is encoded in noun-phrases as pre-modifiers, if placed before the head-noun, and/or as post-modifiers after the head-noun [BC98].

We focus on noun-phrases with adjectives and nouns acting as pre-modifiers (e.g., "the *yellow* cup" and "the *office* door", respectively). We do not address verb phrases acting as pre-modifiers (e.g., "the *swiftly opening* door"), because these refer to activities or events which cannot be handled by our attention models. Furthermore, to avoid in-depth semantic analysis, we ignore post-modifiers which typically are formed by clauses and preposition phrases in noun phrases (e.g., "the author *whose* paper is reviewed" and "the cup *on* the table", respectively).

Parsing

We determine the noun-phrases and their constituents with a shallow parser which is based on regular expressions and was tested on the CoNLL-2000 Corpus [TB00]. Therefore, we trained a Brill tagger, which is backed-off by an n-gram and regular expression tagger, on the Brown corpus [Fra79].

Extraction

Once we have identified the referring noun-phrase and its constituents, we determine the linguistic descriptions that can influence our attention model. First, we match the adjectives against a set of known attributes and their respective linguistic descriptions. Here, we focus on the 11 English basic color terms [BK69].[8] Furthermore, we try to identify references to known object entities. Therefore, we match the object specification (consisting of the pre-modifiers and the head-noun) with a database that stores known object entities and their (exemplary) specifications or names, respectively. We also include adjectives in this matching process, because otherwise semantic analysis is required to handle ambiguous expressions (e.g., "the *Intelligent* Systems Book" or "the *Red* Bull Can"). However, usually either attributes or exact object specification are used, because their combined use is redundant. A major difficulty is that the use of object specifiers varies depending on the user, the conversational context, and the environment. Thus, we have to regard partial specifier matches, e.g."the Hobbits" equals "the Hobbits cookies package". Obviously,

---

[8]Please note that we can easily train color term models for other color term sets (e.g., to integrate additional color terms or to work with different languages) [SF10b, SS12a].

the interpretation of these references depends on the shared conversational context. Given a set of known, possible, or plausible objects (depending on the degree of available knowledge), we can treat this problem with string and tree matching methods by interpreting each specifier as node in a tree. Consequently, we use an edit distance to measure the similarity, see [EIV07], and apply a modified version of the Levenshtein distance that is normalized by the number of directly matching words. Then, we determine the best matching nodes in the tree of known specifications. An object reference is detected, if all nodes in the subtree defined by the best matching node belong to the same object and there do not exist multiple modes with equal minimum distance that belong to different objects.

4.3.2.2 Neuron-Based Saliency Model

To integrate visual saliency, pointing gestures, and spoken target object descriptions, Schauerte and Fink introduced a top-down modulatable saliency model [SF10a]. The model combines Navalpakkam's idea of a modulatable neuron-based model [NI07], which itself is based on ideas of Wolfe et al.'s GSM (see Sect. 2.1.1), with the use of spectral saliency (see Sect. 3.2.1) to calculate the contrast of each neuron's response, see Fig. 4.10. In this model, each feature dimension j—e.g.color, orientation, and lightness—is encoded by a population of N_j neurons with overlapping Gaussian tuning curves (cf. Fig. 2.2) and for each neuron n_{ij} a multi-scale saliency map s_{ij} is calculated. Therefore, we calculate the response $n_{ij}(I^m)$ of each neuron for each scale m of the input image I and use spectral whitening, see Sect. 3.2.1, to calculate the feature maps

(a)

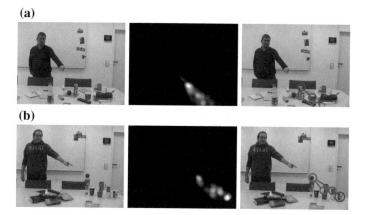

(b)

Fig. 4.10 Example of neuron-based top-down modulation, *left-to-right* the images, their multimodal saliency maps, and the FoA shifts (the initial FoA is marked *blue*). The presented approach reflects how pointing gestures and verbal object references guide the perceptual focus of attention toward the referred-to object of interest. **a** "Give me the Hobbits cookies!", **b** "There is my yellow black cup"

$$s_{ij}^m = g * \mathscr{F}^{-1}\left\{e^{i\Phi\left(\mathscr{F}\{n_{ij}(I^m)\}\right)}\right\}$$ (4.26)

with the Fourier-Transform $\mathscr{F}$, the Phase-Spectrum Φ, and an additional 2D Gaussian filter g. Then, we normalize these single-scale feature maps and use a convex combination in order to obtain the cross-scale saliency map s_{ij}

$$s_{ij} = \sum_{m \in M} w_{ij}^m \mathcal{N}\left(s_{ij}^m\right)$$ (4.27)

with the weights w_{ij}^m and the normalization operator $\mathcal{N}$. The latter performs a cross-scale normalization of the feature map range, attenuates salient activation spots that are caused by local minima of $n_{ij}(I^m)$, and finally amplifies feature maps with prominent activation spots (cf. [IK01b]). However, since we do not incorporate knowledge about the size of the target, we define the weights w_{ij}^m as being uniform, i.e.

$$\sum_{m \in M} w_{ij}^m = 1 .$$ (4.28)

The multi-scale saliency maps s_{ij} of each individual neuron are then combined to obtain the conspicuity maps s_j and the final saliency map S_B

$$s_j = \sum_{i=1}^{N_j} w_{ij} s_{ij} \quad \text{and} \quad S_B = \sum_{i=1}^{N} w_j s_j,$$ (4.29)

given the weights w_j and w_{ij}.

These weights are chosen in order to maximize the signal-to-noise ratio (SNR) between the expected target and distractor saliency (S_T and S_D)

$$\text{SNR} = \frac{\mathbb{E}_{\theta\|T}[S_T]}{\mathbb{E}_{\theta\|D}[S_D]},$$ (4.30)

given known models of the target and distractor features (i.e., $\theta\|T$ and $\theta\|D$). Therefore, we need to predict the SNR for each neuron (i.e., SNR_{ij}) and feature dimension (i.e., SNR_j) to obtain the optimal weights w_{ij} and w_j according to

$$w_{ij} = \frac{\text{SNR}_{ij}}{\frac{1}{n}\sum_{k=1}^{n}\text{SNR}_{kj}} \quad \text{and} \quad w_j = \frac{\text{SNR}_j}{\frac{1}{N}\sum_{k=1}^{N}\text{SNR}_k},$$ (4.31)

as has been proposed by Navalpakkam and Itti [NI06].

The SNR calculation is critical for this model, especially since we aim at using general models for saliency modulation that can also be applied for recognition and naming of objects. This stands in contrast to most prior art, in which saliency modulation was directly learned from target image samples (e.g. [Fri06, IK01b,

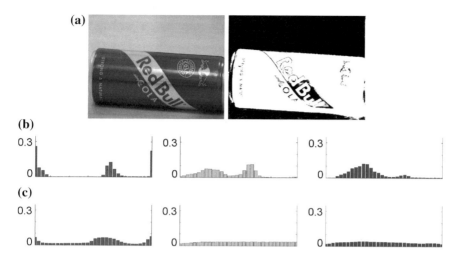

Fig. 4.11 An example of the automatic acquisition of a target object's color models from a model view, see **a** and **b**. For comparison, a combined color term target model for *red* and *blue*, see **c**. **a** Known target object's model view and its segmentation mask obtained with color spatial variance, **b** marginal distributions of the target object's HSL color model, **c** uniform combination of the *red* and *blue color* term models

NI07]; cf. [FRC10]). In our implementation, we use probabilistic target and distractor feature models (i.e., $p(\theta \| T)$ and $p(\theta \| D)$, respectively) and calculate SNR_{ij} and SNR_j according to

$$\text{SNR}_{ij} = \left[\frac{\mathbb{E}_{\theta \| T, I}[s_{ij}]}{\mathbb{E}_{\theta \| D, I}[s_{ij}]} \right]^{\alpha} \quad \text{and} \quad \text{SNR}_j = \frac{\mathbb{E}_{\theta \| T, I}[s_j]}{\mathbb{E}_{\theta \| D, I}[s_j]}, \qquad (4.32)$$

where $\mathbb{E}_{\theta \| T, I}[s_{ij}]$ and $\mathbb{E}_{\theta \| D, I}[s_{ij}]$ is the expected saliency, according to the calculated neuron saliency map s_{ij}, of the respective feature model in image I. Here the constant exponent α is an additional parameter that influences the modulation strength. This is especially useful when dealing with smooth feature models, e.g. color term models (see, e.g., Fig. 4.11), to force a stronger modulation. Analogously, $\mathbb{E}_{\theta \| T, I}[s_j]$ and $\mathbb{E}_{\theta \| D, I}[s_j]$ is the expected target and distractor saliency, respectively, for the calculated conspicuity map s_j, see Eq. 4.29.

4.3.2.3 Target Information and Models

Since our neuron-based saliency model requires to calculate an expected SNR to highlight the target object, we require the target feature information in a probabilistic model. In the following, we explain our three sources of target feature information. First, target color terms (e.g., "red") that we learn from web images. Second, target objects that are generated from images in a target object database, i.e. a set of known objects. Third, information about distracting features such as the color of the background or potential clutter.

Colors

The color models $p(\theta \| T_{color})$, see e.g. Fig. 4.11, are learned using the Google-512 dataset [SF10b], which was gathered from the Internet for the 11 English basic color terms (see Sect. 4.1.1.2). Therefore, we use probabilistic latent semantic analysis with a global background topic and a probabilistic HSL color model [SF10b]. The latter reflects the different characteristics of real-world images and images retrieved from the Internet. Here, we use HSL as color space, because the color channels are decoupled and thus support the use of independent neurons for each channel. However, since color term models are as general as possible, we can in general not expect as strong modulation gains as with specific target object models.

Objects

If we have access to an image of a target object (e.g., the close-up object views that are part of the ReferAT dataset's object database), we can calculate object-specific target feature models $p(\theta \| T_{obj})$. For this purpose, we exploit that the target objects are usually well-centered in the model views and use the color spatial variance—i.e., a known salient object detection feature [LSZ+07]—to perform a foreground/background separation, see Fig. 4.11. Additionally, the acquired segmentation mask is dilated to suppress noise and omit background pixels around the object boundaries. Then, we calculate $p(\theta \| T_{obj})$ as the feature distribution of the foreground image pixels. If we have access to multiple views of the same object, we use a uniform combination to combine the models.

Distractors

In the absence of a pointing gesture, the model of distracting objects and background of each image $p(\theta \| D_I)$ is estimated using the feature distribution of the whole image. Thus, we roughly approximate a background distribution and favor objects with infrequent features. In the presence of a pointing gesture, it is beneficial to incorporate that pointing gestures narrow the spatial domain in which the target object is to be expected. Consequently, we focus the calculation of the distractor feature distribution $p(\theta \| D_I)$ on the spatial region that was indicated by the pointing gesture. Therefore, we calculate a probabilistic map—similarly to the pointing cone, see Eq. 4.12, but with an increased variance σ_c^2—to weight the histogram entries when calculating the feature distribution $p(\theta \| D_I)$. However, since in both cases the target object is part of the considered spatial domain, the distractor feature models are smoothed to avoid suppressing useful target features during the modulation.

4.3.2.4 Conditional Random Field

We rely on the same conditional random field structure that has been explained in Sect. 4.3.1.3 and just adapt the features. Here, we rely on the same target information as the neuron-based approach, see Sect. 4.3.2.3.

Features

One of the major advantages of our machine learning based approach with CRFs is that it is very simple to include additional features and thus target information. To incorporate information about the target object's appearance, we rely on the probabilistic target models that we introduced for the neuron-based approach (see Sect. 4.3.2.3). Therefore, after scaling the image to the CRF's grid size, we calculate the target probability, i.e., $p(\theta \| T_{\text{color}})$ or $p(\theta \| T_{\text{obj}})$, at each image pixel and append the probability to the unary feature vector at the corresponding CRF grid location.

4.3.2.5 The ReferAT Dataset

To evaluate how well multimodal—here, pointing gestures and spoken language—references guide computational attention models, we collected a dataset which contains 242 multimodal referring acts that were performed by 5 persons referring-to a set of 28 objects in a meeting room [SF10a], see Fig. 4.12. This limited set of objects defines a shared context of objects that are plausible in the scene and can be addressed. We chose the objects from a limited set of classes (most importantly: books, cups, packages, and office utensils) with similar intra-class attributes, i.e. size and shape. Thus, in most situations, object names and colors are the most discriminant verbal cues for referring-to the referent. Please note that the limited number of classes further forces the participant to use specifiers to address the objects, because the object class information alone would often lead to ambiguous references.

Hardware Setup

The hardware setup is identical to the PointAT dataset's hardware setup, see Sect. 4.3.1.4. The dataset was recorded using a monocular Sony EVI-D70P pan-tilt-zoom camera. The camera provides images in approximately PAL resolution (762×568 px), and offers an optical zoom of up to $\times 18$. Its wide horizontal opening angle is $48°$. The camera was mounted at eye height of an averagely tall human to reflect a human-like point of view [MFOF08].

Fig. 4.12 Representative object references in our ReferAT evaluation dataset. Depicted are some exemplary images with the corresponding binary target object masks that we can generate based on the annotated target object boundaries

Procedure

We intended to obtain a challenging dataset. Thus, we allowed the participants at every moment to freely change their own position as well as select and arrange the objects that are visible in the scene, see Fig. 4.12. Furthermore, after we explained that our goal is to identify the referent, we even encouraged them to create complex situations. However, naturally the limited field of view of the camera limits the spatial domain, because we did not allow references to objects outside the field of view. Furthermore, we asked the participants to point with their arms extruded, because we use the line-of-sight to estimate pointing direction [RPF08, Ban04] and do not evaluate different methods to determine the pointing direction (cf. [NS07]). In order to verbally refer to an object, the participants were allowed to use arbitrary sentences.

(a)   **(b)**

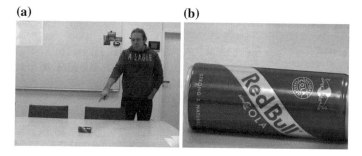

Fig. 4.13 Exemplary acquisition of an object model for the ReferAT dataset's object database. The trainer points to an objects that is then automatically identified (**a**). Then, the system automatically estimates the target object's size and uses the camera's zoom functionality to obtain a close-up image (**b**)

But, since the participants often addressed the object directly, in some cases only a noun phrase was used in order to verbally specify the referent.

We manually transcribed the occurring linguistic references to avoid the influence of speech recognition errors. Furthermore, we annotated the dominant eye, the pointing finger, and the resulting pointing direction. Accordingly, we are able to assess the quality of the automatically recognized pointing gesture and its influence on the detection of the referent. Additionally, for each linguistic reference, we annotated the attributes, target object, and whether the specific target object can be recognized without the visual context of the complementary pointing gesture (e.g. "the cup" vs. "the Christmas elk cup").

Object Database

To make it possible to highlight a known target object as well as to enable object recognition, we collected a database that contains images of all objects that have been referenced in the dataset. For this purpose, we used our automatic pointing reference resolution system in a learning mode (see Sect. 4.3.1.2). We placed each object that had to be learned at a position where the pointing reference was unambiguous. Then, we referred to the object via a pointing gesture and a verbal specification (see Fig. 4.13). Our system then used the pointing gesture to identify the referred-to object, applied segmentation based on maximally stable extremal regions (MSER) [MCUP04] to roughly estimate the object boundaries, and zoom toward the object to obtain a close-up view for learning. These close-up views acquired are stored in a database, in which they are linked with the verbal specification.

Dataset Properties

On average the target object occupies 0.70 % of the image area. Naturally, due to the experimental environment, the target object locations are concentrated in the lower half of the image, the table area. The average differences between the annotated and automatically determined pointing origin and direction are 12.50 px and 3.92°, respectively. Again, the former is caused by the fact that the system is based on the hand's center and not the finger. The latter, again, is due to the fact that the eye positions are estimated given the head-shoulder detection. In some cases, the ray that originates from the finger tip (i.e., pointing origin) and follows the pointed direction does not intersect the target object's boundaries and misses the object. The rate of how often the object's annotated boundary polygon is missed by the pointing ray is 7.85 % for the annotated pointing information and 26.03 % for the automatic detection.

4.3.2.6   Evaluation Results

Procedure and Parameters

To train and evaluate our models, we use a leave-one-person-out training procedure. The CRF is trained with a grid resolution of 381×284 and a 4-connected neighborhood.

As a reference for the neuron-based model, which we have described in detail in Sect. 4.3.2.2, we use the results reported by Schauerte and Fink [SF10a]. Since the performance of the CRF is substantially better than the neuron-based model (the CRF's PHR is often higher than the neuron-based model's FHR), we refrain from (re-)evaluating the neuron-based model to calculate PHR and $\int$ FHR. Schauerte and Fink's model was based on the hue, saturation, lightness, and orientation as feature dimensions. Each feature dimension had a sparse population of 8 neurons. The SNR exponent α was set to 2.

The language processing correctly detected 123 of 123 color references and 123 of 143 references to specific objects (e.g., as negative example: "the tasty Hobbits" as reference to the "the Hobbits cookies package" was not detected; for comparison, as non-trivial positive matching samples, "valensina juice bottle" and "ambient intelligence algorithms book" have been matched to "valensina orange juice package" and "algorithms in ambient intelligence book" in the data base, respectively). Most importantly, the specifier matching of object descriptions made only one critical mismatch ("the statistical elements book" has been matched to "the statistical learning book" instead of "the elements of statistical learning book"). Since wrong target information will lead us to highlight the wrong image areas, this is an important aspect and the reason why we chose such a cautious matching method as described in Sect. 4.3.2.1.

Results

We present the results for different target information conditions in separate tables. Table 4.4 shows the results without any linguistic target information, Table 4.5 contains the results obtained with automatically determined target object information, and Table 4.6 provides the results achieved with groundtruth target information. Each table presents the results achieved without pointing information, with detected pointing information, and with groundtruth pointing information.

First of all, we can notice that the CRFs accurately segmented the target object in most "simple" object arrangements such as in the PointAT dataset, see Fig. 4.9, but they can only predict rough salient regions of interest in complex situations, see Fig. 4.14.

If we use FHR as key evaluation measure—as has been done by Schauerte and Fink [SF10a]—, then the integration of LDRC as a CRF feature clearly improves the results. However, in contrast, the performance as quantified by PHR decreases when LDRC is integrated, at least if pointing information is integrated. This stands in contrast to our results on PointAT and can most likely be explained by the fact that LDRC seems to be unable to highlight a single object in a dense cluster of distractors, which is not surprising given its definition. But, it often highlights small clusters of spatially close objects in which the target object is contained, which increases in FHR while it decreases PHR.

In any case, CRFs clearly outperform the neuron-based model, often by more than a 20 % higher FHR. In fact, the performance of the CRFs with detected pointing information often even outperform the performance of the neuron-based model with groundtruth pointing information, although the use of detected pointing information

Table 4.4 Target object detection performance on the ReferAT dataset without spoken target object information

Algorithm	No language			
	PHR	FHR (%)	∫FHR	NTOS
	No pointing			
Neuron-based	–	9.90	–	–
CRF, no LDRC	6.61 %	18.60	0.307	1.767
CRF, w/LDRC	5.37 %	19.83	0.339	1.808
	Pointing detected			
Neuron-based	–	46.30	–	–
CRF, no LDRC	28.51 %	59.92	0.708	4.565
CRF, w/LDRC	27.69 %	64.05	0.722	4.627
	Pointing annotated			
Neuron-based	–	51.20	–	–
CRF, no LDRC	33.47 %	69.42	0.769	5.511
CRF, w/LDRC	33.06 %	69.83	0.779	5.535

A description of the evaluation measures can be found in Sect. 4.3.1.5

Table 4.5 Target object detection performance on the ReferAT dataset with automatically determined spoken target object information

Algorithm	Detected language			
	PHR	FHR (%)	∫FHR	NTOS
	No pointing			
Neuron-based	–	15.70	–	–
CRF, no LDRC	24.38%	33.88	0.467	3.270
CRF, w/LDRC	24.79%	37.19	0.463	3.318
	Pointing detected			
Neuron-based	–	50.00	–	–
CRF, no LDRC	55.37%	72.73	0.783	6.551
CRF, w/LDRC	52.48%	75.21	0.801	6.497
	Pointing annotated			
Neuron-based	–	63.20	–	–
CRF, no LDRC	66.12%	80.17	0.830	7.377
CRF, w/LDRC	65.29%	81.41	0.834	7.355

A description of the evaluation measures can be found in Sect. 4.3.1.5

Table 4.6 Target object detection performance on the ReferAT dataset with groundtruth spoken target object information

Algorithm	Groundtruth language			
	PHR	FHR (%)	∫FHR	NTOS
	No pointing			
Neuron-based	–	16.50	–	–
CRF, no LDRC	19.83%	30.58	0.424	2.951
CRF, w/LDRC	19.42%	34.71	0.452	3.022
	Pointing detected			
Neuron-based	–	54.10	–	–
CRF, no LDRC	54.13%	72.31	0.780	6.351
CRF, w/LDRC	47.93%	73.55	0.794	6.181
	Pointing annotated			
Neuron-based	–	59.90	–	–
CRF, no LDRC	63.63%	82.23	0.837	7.206
CRF, w/LDRC	60.74%	84.71	0.842	7.069

A description of the evaluation measures can be found in Sect. 4.3.1.5

leads to a substantial drop in the overall performance. The performance difference that is caused by the use groundtruth and detected pointing information can be explained by the imprecision of the detected pointing origin and pointing direction—see the dataset property discussion in Sect. 4.3.2.5—, which often causes the pointing ray to miss the intended target object.

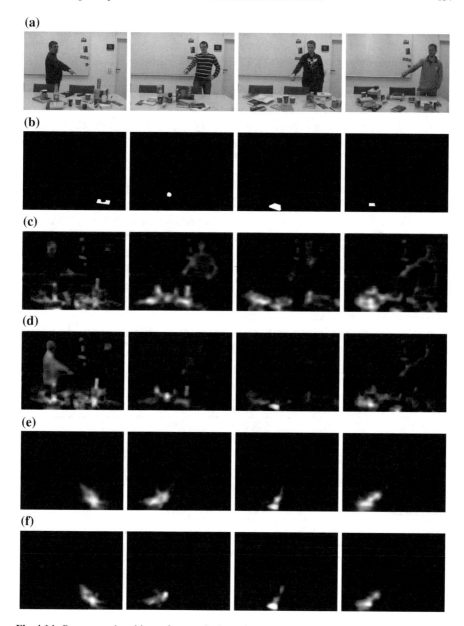

Fig. 4.14 Representative object references in the ReferAT dataset and CRF target region predictions (with groundtruth pointing information and LDRC as additional feature). **a** Images, **b** masks, **c** w/o pointing, w/o language, **d** w/o pointing, with language, **e** with pointing, w/o language, **f** with pointing, with language

Since pointing gestures substantially limit the spatial area in which we expect target objects, it is intuitively clear that the integration of pointing gestures substantially improves the performance under all three target information conditions (i.e., no language, automatically extracted spoken target object information, and annotated target information, see Tables 4.4, 4.5, and 4.6, respectively; compare "no pointing" to "pointing detected" and "pointing annotated").

The integration of language also substantially improves the performance on its own, i.e. without accompanying pointing gestures (compare Tables 4.4, 4.5 and 4.6). Here, we have to differentiate between our two types of target information, i.e. the knowledge of the target object itself or just a color description. To further investigate this aspect, we labeled the color attributes for all objects in the ReferAT dataset. This way, we can train and test CRFs that are always given the correct target object or the correct target object's color term models. As we can see in Table 4.7, if we use the exact target object model, we achieve a better performance compared to the object's color attribute description. This could have been expected, because the color term models are general and not as specific and discriminative as the visual target object models, see Fig. 4.11. Nevertheless, target color term models can guide the attention and lead to substantially better results than models without any verbal target object information (compare Table 4.7 to Table 4.4).

Finally, the combination of both modalities leads to further improvements compared to each unimodal result (see Tables 4.4, 4.5, 4.6, and 4.7) This observation con-

Table 4.7 Target object detection performance on the ReferAT dataset given the groundtruth target object model or the appropriate color term model, see Fig. 4.11

Algorithm	PHR (%)	FHR (%)	$\int$FHR	NTOS
(a) Target object				
	Groundtruth object			
	No pointing			
CRF, no LDRC	25.62	39.26	0.480	3.265
CRF, w/LDRC	24.79	39.67	0.477	3.317
	Pointing annotated			
CRF, no LDRC	71.49	88.84	0.860	7.614
CRF, w/LDRC	70.66	90.50	0.864	7.490
(b) Target color				
	Groundtruth color			
	No pointing			
CRF, no LDRC	21.07	31.82	0.425	3.039
CRF, w/LDRC	19.42	34.30	0.436	3.043
	Pointing annotated			
CRF, no LDRC	60.33	83.88	0.844	6.997
CRF, w/LDRC	59.09	82.23	0.838	6.930

A description of the evaluation measures can be found in Sect. 4.3.1.5

firms that the guidance provided by the individual modalities, i.e. pointing gestures and verbal descriptions, complement each other. This could be expected, because in our scenario both modalities provide different types of information: The pointing gestures guide the attention toward spatial areas of interest, along their spatial corridor of attention. The verbal description provide information about the target object's visual appearance that help to discriminate the object from the background and surrounding clutter.

4.4 Gaze Following in Web Images

In the previous Sect. 4.3, we have focused on a typical HRI task with an evaluation in a laboratory environment. In the final technical section of this book, we will address a topic that we are not even close to solving. However, it represents the logical consolidation of the work that we presented in previous sections. We will show how we can use the methods that we developed to interpret attentional signals in HRI to implement gaze following in web images. In other words, we try to identify the object that is being looked at in images that have been composed by human photographers. Accordingly, we transfer and combine our approaches from previous sections and use conditional random fields with features such as, most importantly, a probabilistic corridor of interest that encodes the gaze direction, spectral saliency detection, and locally debiased region contrast saliency with an explicit center bias.

Internet image collections and datasets pose many challenges compared to data that is acquired in controlled laboratory environments. For example, we have to cope with an extreme variety in the depicted objects and environments, image compositions, and lighting conditions, see Fig. 4.15. The challenges that arise with such

Fig. 4.15 Example Gaze@Flickr images to illustrate the complexity and variance in the dataset

data are also responsible for the fact that even today there does not exist a computer vision algorithm that is able to reliably estimate a gaze direction on the dataset that we present in this section. However, especially the variety of target objects makes it particularly interesting to test our concepts on web images, because our saliency-based approach does not require hundreds or thousands of object detectors to detect all kinds of objects. Furthermore, we are convinced that any approach that we develop on this particularly challenging data will perform even better in simpler scenarios. For example, depth data would undoubtedly assist the identification and segmentation of the target objects, and information about the type and location of potential target objects could serve as valuable prior.

4.4.1 Approach

In principle, we rely on the same methodology as for pointing gestures, see Sect. 4.3.1.

4.4.1.1 Spatial Gaze Target Probability

In principle, gaze serves the same purpose as pointing, i.e. to direct the attention to certain parts of the image. Analogue to Eq. 4.12, we represent the observed gaze direction's attention corridor as

$$p_G(x) = p(\alpha(x, o)|d, o) = \mathcal{N}(0, \sigma_c^2) \,. \tag{4.33}$$

Here, $\alpha(x, o)$ is the angle between the vector from the eyes o to the image point x given the gaze direction d, and σ encodes the assumed gaze direction inaccuracy or uncertainty. This equation represents the probability $p_G(x)$ that the object at point x in the image is being looked-at and defines our probabilistic corridor of attention.

4.4.1.2 Heuristic Integration

Again, see Sect. 4.3.1.2, we implement a heuristic approach to serve as a baseline for our CRFs. Given $p_G(x)$, see Eq. 4.33, we calculate the top-down gaze map

$$S_t(a, b) = p_G(x) \tag{4.34}$$

with $x = (a, b)$. Then, we use QDCT to calculate the visual saliency map S_b based on the PCA decorrelated CIE Lab color space, see Sect. 3.2. The saliency map S_b is normalized to $[0, 1]$. The final heuristic saliency map S is defined as

$$S = S_b \circ S_t \,, \tag{4.35}$$

where $\circ$ represents the Hadarmard product.

4.4.1.3 Conditional Random Field

The CRF relies on the same structure, learning method, and prediction method as in the previous section (Sect. 4.3).

As unary image-based features, we include the following information at each CRF grid point: First, we include each pixel's normalized horizontal and vertical image position in the feature vector. Second, we directly use the pixel's intensity value after scaling the image to the CRF's grid size. Third, we include the scaled probabilistic gaze cone, see Sect. 4.3.1.1. Then, after scaling each saliency map to the appropriate grid size, we append QDCT saliency maps based on the PCA decorrelated Lab color space (see Sect. 3.2.1) at three scales: 96×64 px, 168×128 px, and 256×192 px. Furthermore, we either include the RC'10, LDRC, or LDRC+CB saliency map, see Sect. 4.2.

Again, as CRF edge features, we use a 1-constant and 10 thresholds to encode the color difference of neighboring pixels. Then, we multiply the existing features by an indicator function for each edge type (i.e., vertical and horizontal), which allows to parametrize vertical and horizontal edges separately.

4.4.2 The Gaze@Flickr Dataset

To evaluate the ability to identify at which object a person is looking in web images, we collected a novel dataset that we call Gaze@Flickr. Our dataset itself is based on the MIRFLICKR-1M dataset[9] [HTL10], which consists of 1 million Flickr images under the Creative Commons license. We collected and annotated our dataset in several, subsequent steps:

1. First, we inspected all MIRFLICKR-1M images and selected the images that show at least one person who gazes at something.[10]
2. Then, we selected a subset of 1000 images, for which we outlined the heads/faces of up to three persons.
3. For each annotated head/face region, we annotated the gaze direction under two viewing conditions:

   (a) First, we annotated the gaze direction while being able to see the whole image. We call this the "full" condition.
   (b) Second, we annotated the gaze direction while only being shown the face/head region. For this purpose, the other parts of the image were shown as being black. We call this the "blank" condition.

---

[9]http://press.liacs.nl/mirflickr/.

[10]During this process, we also collected all images that depict persons pointing at something. However, we could not find a sufficient number of such images to build a "pointing gestures in the wild dataset".

Fig. 4.16 Representative object references in our Gaze@Flickr dataset. Depicted are some exemplary images with their corresponding binary target object masks that we can generate based on the annotated target object boundaries

4. Then, for each face/head with an annotated gaze direction, we annotated the boundary of the object at which the person is looking, see Fig. 4.16. In some cases the target object was either "ambiguous" (i.e., the target was not visible or there were several equally plausible target objects) or—most likely—"outside" the image (i.e., not depicted in the image). In both cases, we were unable to label the target object and instead just tagged the images accordingly.

In total, our dataset contains 863 images that depict 1221 annotated gaze references, excluding gaze samples with unclear or out-of-sight targets.

Properties

On average the target object occupies 4.33 % of the image area, which makes the objects substantially larger compared to the target objects in the PointAT (0.58 %) and ReferAT (0.70 %) datasets. However, at the same time, the objects are also considerably smaller compared to the MSRA dataset (18.25 %). As could be expected, the gaze annotation viewing condition (i.e., full or blank) influences the annotated gaze directions. The average difference between the annotated directions is $12.10°$. Similar to pointing gestures, the ray that originates from the eyes (i.e., gaze origin) and follows the gaze direction can miss the target object's polygon. The rate of how often the object's annotated target polygon is missed depends substantially on the viewing condition: The rays that were annotated under the full condition miss only 4.67% of the target objects, while the rays under the blank conditions miss 26.94% of the targets, i.e. $5.76\times$ more often. This, in combination with the substantial deviation of annotated gaze directions, demonstrates the important influence that context information—i.e., the information about potential target objects—can have on gaze estimates made by humans.

4.4.3 Evaluation

4.4.3.1 Procedure and Parameters

To train and evaluate our CRFs, we follow a 5-fold cross-validitation procedure. Accordingly, we have about 976 training images and 244 test images for each fold. The CRF is trained with a grid resolution of 300×300 and a 4-connected neighborhood.

We have to rely on the two annotated gaze directions in the evaluation. This is due to the fact that there does not exist a computer vision method that is able to reliably produce gaze estimates on our Gaze@Flickr dataset. Accordingly, since we can not estimate a gaze direction uncertainty or inaccuracy, we use a fixed probabilistic gaze cone σ of approximately $14°$, i.e. 0.25 rad.

4.4.3.2 Measures

We use all evaluation measures for salient object detection and focus of attention selection that we have used in Sects. 4.4.3.2 and 4.3.1.5, respectively.

4.4.3.3 Data Analysis

Do People Look at Salient Things?

An interesting question is whether or not the objects that people are looking at are already perceptually salient (the opposite question would be "Do people in images frequently look at objects that do not pop-out from the background?"). We can use the NTOS evaluation measure to investigate this question, see Sect. 4.3.1.5. NTOS compares the mean saliency at target object location to the mean saliency of the background. The measure is normalized in terms of the standard deviation of the saliency values. Thus, NTOS ≤ 0 means that the saliency at the target object's location is not higher compared to the background, whereas an NTOS of 1 means that the saliency in the target object area is one standard deviation above average.

For this purpose, we calculated the visual saliency of several algorithms without gaze integration, see Table 4.8. Apparently, the NTOS is substantially higher than 0 for almost all evaluated visual saliency algorithms. This means that people in photographs often look at objects that are perceptually salient. It is in fact very interesting that visual saliency algorithms seem so capable to highlight the target objects, because it indicates that the images are composed in a way that the gaze of persons that view the image is likely to be attracted by the target image regions (see Sect. 3.2). The fact that AC'09 has a negative NTOS can be explained by the fact that the target objects are too small and non-target areas too heterogeneous for Achanta's simple approach [AHES09].

Are the Target Locations Center Biased?

Since Gaze@Flickr is composed of web images, we can expect that the data is influenced by photographer biases (see Sect. 4.2). If we look at the results in Table 4.8, we can answer this question without an empirical analysis. Without integrated gaze information, LDRC+CB performs better than LDRC with respect to almost all evaluation measures. This indicates that the data is center biased. However, the target object location radii do not follow a Gaussian distribution, see Fig. 4.18. In fact, we can reject the hypothesis of a Gaussian distribution with separate tests such as, e.g., the Jarque-Bera test ($p = 0.001$). This could be expected, because in a considerable number of images the gazing person is in the image's center and not the target object. Nonetheless, our Gaussian center bias achieves a better performance than RC'10's intrinsic bias, see Table 4.8.

4.4.3.4   Results

Our quantitative evaluation results are shown in Table 4.8 and example CRF predictions are depicted in Fig. 4.17.

We would like to start with a short reminder that we rely on two classes of evaluation measures: First, measures that evaluate the ability to focus the target object (PHR, FHR, and $\int$FHR). Second, measures that mainly evaluate the saliency map's ability to separate the target object from the background (NTOS, F_β, F_1, and $\int$ROC). At this point, we would like to explain one of the reasons, why we did not employ the F_β and F_1 measure in Sect. 4.3. As we can see in Table 4.8, the F_β and F_1 values are substantially smaller compared to the values that we observed on the MSRA dataset, see Table 4.1. This is related to the smaller target object size, because the errors made for the background pixels have a stronger influence on the evaluation measure for smaller target object sizes. If we consider that the target objects in the PointAT and ReferAT dataset are even smaller, it is understandable that the evaluation measures lose their informative value on these datasets.

Table 4.8 Target object detection performance on the Gaze@Flickr dataset

Method	PHR (%)	FHR (%)	$\int$FHR	NTOS	F_β	F_1	$\int$ROC
	No gaze						
Center bias	8.44	12.37	0.1531	0.6231	0.0664	0.0948	0.5959
Heuristic, CCH'12	12.20	15.48	0.3201	0.7212	0.0844	0.1063	0.6812
Heuristic, GBVS'07	13.92	19.08	0.2209	0.8983	0.0642	0.0920	0.5775
Heuristic, IK'98	13.35	18.26	0.2140	0.8711	0.0684	0.0974	0.6065
Heuristic, PFT'07	11.71	15.15	0.3136	0.8240	0.0856	0.1059	0.7061
Heuristic, DCT'11	12.29	15.89	0.3167	0.8440	0.0802	0.1035	0.6782
Heuristic, EPQFT	11.06	14.17	0.3223	0.8369	0.0869	0.1090	0.7079
Heuristic, QDCT	11.63	15.23	0.3256	0.8625	0.0833	0.1066	0.6905
Heuristic, AC'09	2.54	3.44	0.0973	-0.1437	0.0480	0.0656	0.4575
Heuristic, MSSS'10	10.48	13.60	0.2534	0.5087	0.0762	0.0969	0.6215
Heuristic, RC'10	12.29	17.12	0.2434	0.6270	0.0774	0.1002	0.6603
Heuristic, LDRC	11.06	14.91	0.2248	0.5354	0.0706	0.0930	0.6463
Heuristic, LDRC+CB	13.35	17.94	0.2501	0.7180	0.0775	0.1063	0.6569
Heuristic, QDCT+LDRC+CB	11.47	14.99	0.3321	0.9291	0.0848	0.1097	0.6993
CRF, QDCT	8.51	14.94	0.2079	0.7330	0.0815	0.0890	0.5583
CRF, QDCT & RC'10	11.08	19.98	0.2741	0.9204	0.1019	0.1188	0.6215
CRF, QDCT & LDRC	9.79	16.12	0.2319	0.8158	0.0898	0.1045	0.5945
CRF, QDCT & LDRC+CB	14.34	22.55	0.3116	0.9851	0.1116	0.1285	0.6608
	Gaze "blank"						
Heuristic, QDCT	28.91	38.17	0.5634	1.9352	0.1234	0.1561	0.7590
Heuristic, RC'10	24.24	34.23	0.5028	1.6816	0.1212	0.1526	0.7580
Heuristic, LDRC	24.08	33.66	0.4868	1.5990	0.1181	0.1514	0.7668
Heuristic, LDRC+CB	24.16	34.64	0.5013	1.6915	0.1265	0.1632	0.7573
Heuristic, QDCT+LDRC+CB	29.89	39.48	0.5956	2.0258	0.1264	0.1603	0.7619
CRF, QDCT	29.48	45.80	0.5228	1.8771	0.2053	0.2314	0.8068
CRF, QDCT & RC'10	33.63	49.75	0.5709	1.9805	0.2108	0.2387	0.8216
CRF, QDCT & LDRC	31.75	48.76	0.5644	1.9703	0.2121	0.2381	0.8180
CRF, QDCT & LDRC+CB	34.72	52.62	0.6053	2.0509	0.2226	0.2466	0.8303
	Gaze "full"						
Heuristic, QDCT	32.76	42.10	0.6303	2.2066	0.1314	0.1655	0.7715
Heuristic, RC'10	31.20	42.75	0.5971	1.9711	0.1317	0.1650	0.7739
Heuristic, LDRC	31.29	42.26	0.5817	1.8760	0.1308	0.1618	0.7858
Heuristic, LDRC+CB	30.96	42.59	0.5816	1.9951	0.1388	0.1782	0.7741
Heuristic, QDCT+LDRC+CB	34.64	45.37	0.6715	2.3198	0.1350	0.1705	0.7741
CRF, QDCT	40.06	57.67	0.6572	2.5321	0.2579	0.2923	0.8887
CRF, QDCT & RC'10	43.62	65.18	0.7114	2.6589	0.2652	0.2970	0.8932
CRF, QDCT & LDRC	43.62	62.61	0.6957	2.5936	0.2628	0.2963	0.8903
CRF, QDCT & LDRC+CB	44.02	66.17	0.7263	2.7071	0.2712	0.3024	0.8981

A description of the evaluation measures can be found in Sect. 4.3.1.5

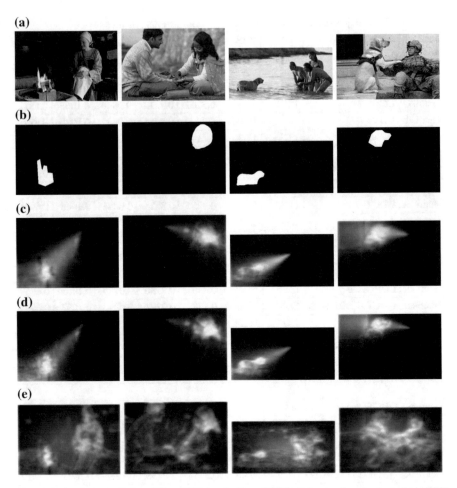

Fig. 4.17 Example predictions of our CRFs on the Gaze@Flickr dataset. **a** Images, **b** masks, **c** CRF with QDCT & LDRC+CB, **d** CRF with QDCT & LDRC, **e** CRF with QDCT & LDRC, no gaze information

Without integrated gaze information and without CRFs, we can see that QDCT exhibits a better salient object detection performance (i.e., F_β, F_1, and $\int$ROC) than the region contrast algorithms (i.e., LDRC, LDRC+CB, and RC'10), whereas LDRC+CB and RC'10 provide a better performance in terms of PHR and FHR. Here, we can observe the influence of the center bias, because LDRC exhibits a lower performance than RC'10, LDRC+CB and, as a sidenote, also QDCT. If we integrate gaze, it is evident that QDCT is superior in terms of all evaluation measures that are related to the FoA on the basis of the "blank" gaze annotation. Furthermore, LDRC+CB provides a better performance with the "full" gaze annotation in terms of salient object detection evaluation measures. However, this situation is not perfectly consistent over both gaze annotations.

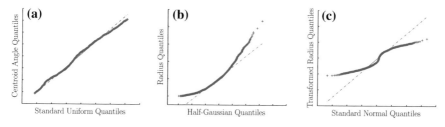

Fig. 4.18 Quantile-quantile (Q-Q) plots of the angles versus a uniform distribution (**a**), radii versus a half-Gaussian distribution (**b**), transformed radii (see Sect. 4.2.2.2) versus a normal distribution (**c**). Compare to the Q-Q plots on MSRA in Fig. 4.4

Similar to our experience with pointing gestures and spoken target information (see Sects. 4.3.1 and 4.3.2; "detected" vs. "annotated"), we can observe that the evaluated CRFs are often able to provide a better performance with the "blank" gaze annotation than the heuristic baselines with the "full" gaze direction, especially in terms of salient object detection (i.e., F_β, F_1, and $\int$ROC). This comes at no surprise, since CRF are well known for their good performance in various image segmentation tasks, see Sect. 4.1.2.1. If we compare the CRFs to the heuristic method with the same gaze annotation, we can see that the CRFs outperform the heuristic gaze integration by a considerable margin, see Table 4.8.

Since we want to integrate our work on salient object detection and visual saliency, we investigate the performance that we can achieve when we let the CRFs combine QDCT with LDRC, LDRC+CB, or RC'10. To serve as a heuristic baseline, we present the results that we can achieve with a linear integration of the saliency maps of QDCT and LDRC+CB (i.e., "QDCT+LDRC+CB"). As can be seen in Table 4.8, the CRFs that use the saliency maps of QDCT and LDRC+CB as features are able to achieve a considerably higher performance than the heuristic linear integration scheme. If we compare the quantitative results of the evaluated CRFs, then LDRC+CB is the basis for the best performance, followed by RC'10, and then LDRC; almost perfectly consistent over all evaluation measures. The fact that LDRC+CB and RC'10 lead to a better performance than LDRC is not surprising, because the target objects in the Gaze@Flickr dataset appear to be biased toward the image center, as has been discussed in Sect. 4.4.3.3. However, the radii do not follow half-Gaussian distribution, as is apparent in Fig. 4.18. Nonetheless, our explicit Gaussian center bias model[11] in the LDRC+CB algorithm leads to a better performance compared to RC'10's intrinsic bias. Accordingly, we can assume that the Gaussian bias better reflects the actual target object location distribution than the intrinsic bias of RC'10.

---

[11] Please note that we could have replaced the Gaussian center bias model with a model specifically adapted to the Gaze@Flickr dataset. We refrained from doing so, because we prefer not to overadapt to the Gaze@Flickr dataset.

4.4.3.5   Future Work

Unlike in the other technical sections of this book, we would like to end with an outlook on future work, because we described exploratory work that leaves many aspects open. In many images in the Gaze@Flickr dataset, people look at people or faces. Accordingly, we would like to integrate face and person detection as another feature for the CRF. We actually refrained from doing so in the presented evaluation, because of the evaluation's focus on our integration of the aspects described in Sects. 4.2 and 4.3. Furthermore, we have performed experiments with additional features such as, most importantly, histogram of oriented gradients (HOG) and local binary patterns (LBP). These features efficiently encode information about edges and segments around each image pixel and thus can further improve the performance. However, we are currently unable to train the CRFs with these additional features for the complete Gaze@Flickr dataset, because the training would require more random access memory (RAM) than is available on our servers. As a sidenote, we require roughly 110 GB of memory to train the models that achieve the results presented in Table 4.8. Finally and most importantly, we would like to replace the groundtruth that currently forms the foundation of our experiments with automatically estimated gaze directions. However, since gaze estimation in the wild is still an unsolved problem, there currently does not exist any computer vision algorithm that is able to provide gaze estimates on our Gaze@Flickr dataset. Consequently, we are also interested in using the Gaze@Flickr dataset to develop novel gaze estimation methods that work on images that we can find in the web and other largely unconstrained image sources.

4.5   Summary and Future Directions

We presented how we can identify an object of interest that another person wants us to look at and analyze. We addressed two domains: First, attentional signals in human-robot interaction (HRI) that can guide the perceptual saliency. Second, images that have been composed by photographers; without and with the display of additional attentional signals such as people looking at things. For this purpose we came in contact with a wide range of research fields such as salient object detection, photographic image composition, perceptual saliency, image segmentation, HRI, pointing gestures, natural language processing and dialogue, and joint attention.

We initially addressed visual saliency models for HRI in 2010. At this point, only individual relevant aspects have been addressed, most importantly: First, the inherent inaccuracy of pointing gestures and the concept of a corridor of attention. Second, that the perceptual saliency can influence the generation and resolution of multimodel referring acts. Third, how linguistic references and knowledge about a target object's visual appearance can influence visual search patterns. However, there did not exist any computer vision or robotic system that systematically integrated these aspects. Furthermore, all systems that tried to identify pointed-at objects relied on the assumption that all objects and their locations in the environment are known beforehand. Among other aspects, this also contradicted one important goal in robotics: the ability to being able to intuitively teach a robot about unknown things. Accordingly, we integrated several previously isolated ideas, methods, and models to guide the visual saliency and this way help to establish joint attention in multimodal HRI. Here, we are able to efficiently guide the focus of attention in multimodal HRI toward image locations that are highly likely to depict the referent; often being able to directly identify the intended target object.

We knew from our preceding work on eye fixation prediction about the importance of dataset biases such as and most importantly the photographer bias (Sect. 3.2). Thus, we were surprised to find that this aspect has not been addressed for salient object detection, although the bias was apparent in the datasets. Since we were interested to apply salient object detection for our HRI tasks, we started to analyze, model, and remove the center bias in salient object detection methods to facilitate the transfer of such methods to other data domains. This way, we improved the state-of-the-art in salient object detection and derived the currently best unbiased salient object detection algorithm.

After having studied how photographers compose web images and people use attentional signals in interaction to direct our attention, we became interested in the combination of both research directions. Therefore, we collected a novel dataset to learn to identify the object that is being looked at in web images. In this type of images the photographer's placement of objects as well as visible attentional signals such as gaze direct our attention toward specific objects that form a central element of these images. Thus, similar to pointing gestures, we address a different problem than almost all work on gaze estimation, i.e. we are not interested in an exact gaze direction estimates and instead focus on the identification and segmentation of the image area that is being looked at. Again, with no knowledge about the target object's visual appearance, class, type, size, or any other identifying information – except that it is being looked at. Here, we have to deal with several challenges, ranging from the huge image data variance to the sheer non-existence of reliable gaze estimation methods for this type of unconstrained image data. We have achieved promising results with the methods that we initially developed for HRI data. Yet, we consider this work as being mainly exploratory to help us assess the potential of our methods in unconstrained environments with huge deviations in, for example, illumination, depicted content, and image composition. Consequently, there remain many challenges and aspects of future work.

Future Work

Related to identifying the salient object in web images, we see a lot of potential in the integration of descriptors that encode the coarse layout, gist, or global context of the scene, because this information is typically related to the general image composition (e.g., does the image focus on one object? Is there a visible horizon? Is it indoors or outdoors?). Since eye tracking experiments have already been an important aspect of our work (see Sect. 3.2), a logical next step would be to evaluate how good our models predict human gaze patterns for images and videos in which persons use verbal and/or non-verbal attentional signals. Since such datasets do no exist yet, an important aspect would be to either extend our existing datasets with eye tracking data or to create a new dataset specifically for eye tracking studies. With respect to gaze estimation in the wild, we have observed that human gaze direction estimates depend on the image content and context. Accordingly, as an important aspect of our future work, we want to jointly estimate the gaze direction and the image region that is being looked at.

References

[AHES09]   Achanta, R., Hemami, S., Estrada, F., Süsstrunk, S.: Frequency-tuned salient region detection. In: Proceedings of International Conference on Computer Vision Pattern Recognition (2009)

[AS10]   Achanta, R., Süsstrunk, S.: Saliency detection using maximum symmetric surround. In: Proceedings of International Conference on Image Processing (2010)

[ADF10]   Alexe, B., Deselaers, T., Ferrari, V.: What is an object? In: Proceedings of International Conference on Computer Vision Pattern Recognition, pp. 73–80 (2010)

[Ban04]   Bangerter, A.: Using pointing and describing to achieve joint focus of attention in dialogue. Psychol. Sci. 15(6), 415–419 (2004)

[BO06]   Bangerter, A., Oppenheimer, D.M.: Accuracy in detecting referents of pointing gestures unaccompanied by language. Gesture 6(1), 85–102 (2006)

[BP94]   Baluja, S., Pomerleau, D.: Non-intrusive gaze tracking using artificial neural networks. In: Advances in Neural Information Processing Systems (1994)

[Bat81]   Batschelet, E.: Circular statistics in biology 24(4) (1981)

[BK11]   Begum, M., Karray, F.: Integrating visual exploration and visual search in robotic visual attention: the role of human-robot interaction. In: Proceedings of International Conference on Robotic Automation (2011)

[BJ80]   Bera, A.K., Jarque, C.M.: Efficient tests for normality, homoscedasticity and serial independence of regression residuals. Econ. Lett. 5(3), 255–259 (1980)

[BK69]   Berlin, B., Kay, P.: Basic Color Terms: Their Universality and Evolution. University of California Press (1969)

[BC98]   Beun, R., Cremers, A.: Object reference in a shared domain of conversation. Pragmat. Cognit. 1(6), 111–142 (1998)

[BDT08]   Bock, S., Dicke, P., Thier, P.: How precise is gaze following in humans? Vis. Res. 48, 946–957 (2008)

[BSI12]   Borji, A., Sihite, D.N., Itti, L.: Salient object detection: a benchmark. In: Proceedings of European Conference on Computer Vision (2012)

[Bri95]   Brill, E.: Transformation-based error-driven learning and natural language processing: a case study in part-of-speech tagging. Comp. Ling. 21(4), 543–565 (1995)

[Bro07]   Brooks, A.G.: Coordinating human-robot communication. Ph.D. Dissertation, MIT (2007)

[Bus35]   Busswell, G.T.: How People Look at Pictures: A Study of the Psychology of Perception in Art. University of Chicago Press (1935)

[BI00]   Butterworth, G., Itakura, S.: How the eyes, head and hand serve definite reference. Br. J. Dev. Psychol. 18, 25–50 (2000)

[CC03]   Cashon, C., Cohen, L.: The Construction, Deconstruction, and Reconstruction of Infant Face Perception. NOVA Science Publishers (2003). ch. The development of face processing in infancy and early childhood: current perspectives pp. 55–68

[CZM+11]   Cheng, M.M., Zhang, G.-X., Mitra, N.J., Huang, X., Hu, S.-M.: Global contrast based salient region detection. In: Proceedings of International Conference on Computer Vision Pattern Recognition (2011)

[CSB83]   Clark, H.H., Schreuder, R., Buttrick, S.: Common ground and the understanding of demonstrative reference. J. Verb. Learn. Verb. Behav. **22**, 245–258 (1983)

[Dom13]   Domke, J.: Learning graphical model parameters with approximate marginal inference. IEEE Trans. Pattern Anal. Mach. Intell. **35**(10), 2454–2467 (2013)

[DSS06]   Doniec, M., Sun, G., Scassellati, B.: Active learning of joint attention. Humanoids (2006)

[DSHB11]   Droeschel, D., Stückler, J., Holz, D., Behnke, S.: Towards joint attention for a domestic service robot—person awareness and gesture recognition using time-of-flight cameras. In: Proceedings of International Conference on Robotics and Automation (2011)

[ESP08]   Einhäuser, W., Spain, M., Perona, P.: Objects predict fixations better than early saliency. J. Vis. **8**(14) (2008)

[EI08]   Elazary, L., Itti, L.: Interesting objects are visually salient. J. Vis. **8**(3), 1–15 (2008)

[EIV07]   Elmagarmid, A., Ipeirotis, P., Verykios, V.: Duplicate record detection: a survey. IEEE Trans. Knowl. Data Eng. **19**(1), 1–16 (2007)

[FH04]   Felzenszwalb, P.F., Huttenlocher, D.P.: Efficient graph-based image segmentation. Int. J. Comput. Vis. **59**(2), 167–181 (2004)

[FBH+08]   Foster, M.E., Bard, E.G., Hill, R.L., Guhe, M., Oberlander, J., Knoll, A.: The roles of haptic-ostensive referring expressions in cooperative, task-based human-robot dialogue. In: Proceedings of International Conference on Human-Robot Interaction, pp. 295–302 (2008)

[Fra79]   Francis, W.N., Kucera, H.: Compiled by, "A standard corpus of present-day edited american english, for use with digital computers (brown)", 1964, 1971, 1979

[Fri06]   Frintrop, S.: VOCUS: A Visual Attention System for Object Detection and Goal-Directed Search, ser. Springer, Lecture Notes in Computer Science (2006)

[FBR05]   Frintrop, S., Backer, G., Rome, E.: Selecting what is important: training visual attention. In: Proceedings of KI (2005)

[FRC10]   Frintrop, S., Rome, E., Christensen, H.I.: Computational visual attention systems and their cognitive foundation: a survey. ACM Trans. Appl. Percept. **7**(1), 6:1–6:39 (2010)

[GRK07]   Gergle, D., Rosé, C.P., Kraut, R.E.: Modeling the impact of shared visual information on collaborative reference. In: Proceedings of International Conference on Human Factors Computing Systems (CHI), pp. 1543–1552 (2007)

[GZMT12]   Goferman, S., Zelnik-Manor, L., Tal, A.: Context-aware saliency detection. IEEE Trans. Pattern Anal. Mach, Intell (2012)

[GMZ08]   Guo, C., Ma, Q., Zhang, L.: Spatio-temporal saliency detection using phase spectrum of quaternion fourier transform. In: Proceedings of International Conference on Computer Vision Pattern Recognition (2008)

[GZ10]   Guo, C., Zhang, L.: A novel multiresolution spatiotemporal saliency detection model and its applications in image and video compression. IEEE Trans. Image Process. **19**, 185–198 (2010)

[HJ10]   Hansen, D.W., Ji, Q.: In the eye of the beholder: a survey of models for eyes and gaze. Trans. PAMI **32**, 478–500 (2010)

[HKP07]   Harel, J., Koch, C., Perona, P.: Graph-based visual saliency. In: Proceedings of Annual Conference on Neural Information Processing Systems (2007)

[HSK+10]   Hato, Y., Satake, S., Kanda, T., Imai, M., Hagita, N.: Pointing to space: modeling of deictic interaction referring to regions. In: Proceedings of International Conference on Human-Robot Interaction, pp. 301–308 (2010)

[HRB+04]   Heidemann, G., Rae, R., Bekel, H., Bax, I., Ritter, H.: Integrating context-free and context-dependent attentional mechanisms for gestural object reference. Mach. Vis. Appl. **16**(1), 64–73 (2004)

[Hob05]   Hobson, R.: Joint Attention: Communication and Other Minds. Oxford University Press (2005). ch. What puts the jointness in joint attention?, pp. 185–204

[HRM11]   Huettig, F., Rommers, J., Meyer, A.S.: Using the visual world paradigm to study language processing: a review and critical evaluation. Acta Psychologica **137**(2), 151–171 (2011)

[HTL10]   Huiskes, M.J., Thomee, B., Lew, M.S.: New trends and ideas in visual concept detection. In: ACM International Conference on Multimedia Information Retrieval (2010)

[IK01b]   Itti, L., Koch, C.: Feature combination strategies for saliency-based visual attention systems. J. Electr. Imaging **10**(1), 161–169 (2001)

[IKN98]   Itti, L., Koch, C., Niebur, E.: A model of saliency-based visual attention for rapid scene analysis. IEEE Trans. Pattern Anal. Mach. Intell. **20**(11), 1254–1259 (1998)

[JWY+11]   Jiang, H., Wang, J., Yuan, Z., Liu, T., Zheng, N.: Automatic salient object segmentation based on context and shape prior. In: Proceedings of British Conference on Computer Vision (2011)

[KH06]   Kaplan, F., Hafner, V.: The challenges of joint attention. Interact. Stud. **7**(2), 135–169 (2006)

[KF11]   Klein, D.A., Frintrop, S.: Center-surround divergence of feature statistics for salient object detection. In: Proceedings of International Conference on Computer Vision (2011)

[KC06]   Knoeferle, P., Crocker, M.W.: The coordinated interplay of scene, utterance, and world knowledge: evidence from eye tracking. Cognit. Sci. **30**, 481–529 (2006)

[KLP+06]   Kranstedt, A., Lücking, A., Pfeiffer, T., Rieser, H., Wachsmuth, I.: Deixis: how to determine demonstrated objects using a pointing cone. In: Proceedings of the International Gesture Workshop, vol. 3881 (2006)

[KCP+13]   Krause, E.A., Cantrell, R., Potapova, E., Zillich, M., Scheutz, M.: Incrementally biasing visual search using natural language input. In: Proceedings of International Conference on Autonomous Agents and Multi-Agent Systems (2013)

[LMP01]   Lafferty, J.D., McCallum, A., Pereira, F.C.N.: Conditional random fields: Probabilistic models for segmenting and labeling sequence data. In: Proceedings of International Conference on Machine Learning (2001)

[Lil67]   Lilliefors, H.W.: On the Kolmogorov-Smirnov test for normality with mean and variance unknown. J. Am. Stat. Assoc. **62**(318), 399–402 (1967)

[LSZ+07]   Liu, T., Sun, J., Zheng, N.N., Tang, X., Shum, H.-Y.: Learning to detect a salient object. In: Proceedings of International Conference on Computer Vision Pattern Recognition (2007)

[LB05]   Louwerse, M., Bangerter, A.: Focusing attention with deictic gestures and linguistic expressions. In: Proceedings of Annual Conference of the Cognitive Science Society (2005)

[MCUP04]   Matas, J., Chum, O., Urban, M., Pajdla, T.: Robust wide-baseline stereo from maximally stable extremal regions. Image Vis. Comput. **22**(10), 761–767 (2004)

[MFOF08]   McDowell, M.A., Fryar, C.D., Ogden, C.L., Flegal, K.M.: Anthropometric reference data for children and adults: United states, 2003–2006. Technical Report, National Health Statistics Reports (2008)

[MON08]   Minock, M., Olofsson, P., Näslund, A.: Towards building robust natural language interfaces to databases. In: Proceedings of International Conference on Applied Natural Language Information Systems, pp. 187–198 (2008)

[Moj05]   Mojsilovic, A.: A computational model for color naming and describing color composition of images. IEEE Trans. Image Process. **14**(5), 690–699 (2005)

[MN07b]   Mundy, P., Newell, L.: Attention, joint attention, and social cognition. Curr. Dir. Psychol. Sci. **16**(5), 269–274 (2007)

[NHMA03]   Nagai, Y., Hosoda, K., Morita, A., Asada, M.: A constructive model for the development of joint attention. Connect. Sci. **15**(4), 211–229 (2003)

[NI06]   Navalpakkam, V., Itti, L.: An integrated model of top-down and bottom-up attention for optimizing detection speed. In: Proceedings of International Conference on Computer Vision Pattern Recognition (2006)

[NI07]   Navalpakkam, V., Itti, L.: Search goal tunes visual features optimally. Neuron **53**(4), 605–617 (2007)

[NS07]   Nickel, K., Stiefelhagen, R.: Visual recognition of pointing gestures for human-robot interaction. Image Vis. Comput. **25**(12), 1875–1884 (2007)

[NIS12]   NIST/SEMATECH: Engineering Statistics Handbook (2012)
[PN03]   Parkhurst, D., Niebur, E.: Scene content selected by active vision. Spatial Vis. **16**(2), 125–154 (2003)
[PLN02]   Parkhurst, D., Law, K., Niebur, E.: Modeling the role of salience in the allocation of overt visual attention. Vis. Res. **42**(1), 107–123 (2002)
[PPI09]   Passino, G., Patras, I., Izquierdo, E.: Latent semantics local distribution for crf-based image semantic segmentation. In: Proceedings of British Conference on Computer Vision (2009)
[Pea00]   Pearson, K.: On the criterion that a given system of deviations from the probable in the case of a correlated system of variables is such that it can be reasonably supposed to have arisen from random sampling. Philos. Mag. **50**(302), 157–175 (1900)
[PIIK05]   Peters, R., Iyer, A., Itti, L., Koch, C.: Components of bottom-up gaze allocation in natural images. Vis. Res. **45**(18), 2397–2416 (2005)
[Pet03]   Peterson, B.F.: Learning to See Creatively. Amphoto Press (2003)
[Piw07]   Piwek, P.L.A.: Modality choice for generation of referring acts: pointing versus describing. In: Proceedings of International Workshop on Multimodal Output Generation (2007)
[RZ99]   Reinagel, P.. Zador, A.M.: Natural scene statistics at the centre of gaze. In: Network: Computation in Neural Systems, pp. 341–350 (1999)
[RPF08]   Richarz, J., Plötz, T., Fink, G.A.: Real-time detection and interpretation of 3D deictic gestures for interaction with an intelligent environment. In: Proceedings of International Conference on Pattern Recognition, pp. 1–4 (2008)
[Roy05]   Roy, D.: Grounding words in perception and action: computational insights. Trends Cognit. Sci. **9**(8), 389–396 (2005)
[RVES10]   Rybok, L., Voit, M., Ekenel, H.K., Stiefelhagen, R.: Multi-view based estimation of human upper-body orientation. In: Proceedings of International Conference on Pattern Recognition (2010)
[SYH07]   Sato, E., Yamaguchi, T., Harashima, F.: Natural interface using pointing behavior for human-robot gestural interaction. IEEE Trans. Ind. Electron. **54**(2), 1105–1112 (2007)
[Sch14]   Schauerte, B.: Multimodal computational attention for scene understanding. Ph.D. Dissertation, Karlsruhe Institute of Technology (2014)
[SF10a]   Schauerte, B., Fink, G.A.: Focusing computational visual attention in multi-modal human-robot interaction. In: Proceedings of 12th International Conference on Multimodal Interfaces and 7th Workshop on Machine Learning for Multimodal Interaction (ICMI-MLMI). ACM, Beijing (2010)
[SF10b]   Schauerte, B., Fink, G.A.: Web-based learning of naturalized color models for human-machine interaction. In: Proceedings of 12th International Conference on Digital Image Computing: Techniques and Applications (DICTA). IEEE, Sydney (2010)
[SS12a]   Schauerte, B., Stiefelhagen, R.: Learning robust color name models from web images. In: Proceedings of 21st International Conference on Pattern Recognition (ICPR). IEEE, Tsukuba (2012)
[SS13a]   Schauerte, B., Stiefelhagen, R.: How the distribution of salient objects in images influences salient object detection. In: Proceedings of 20th International Conference on Image Processing (ICIP). IEEE, Melbourne (2013)
[SRF10]   Schauerte, R., Richarz, J., Fink, G.A.: Saliency-based identification and recognition of pointed-at objects. In: Proceedings of 23rd International Conference on Intelligent Robots and Systems (IROS). IEEE/RSJ, Taipei (2010)
[SHH+08]   Schmidt, J., Hofemann, N., Haasch, A., Fritsch, J., Sagerer, G.: Interacting with a mobile robot: evaluating gestural object references. In: Proceedings of International Conference on Intelligent Robots and Systems, pp. 3804–3809 (2008)
[SW65]   Shapiro, S.S., Wilk, M.B.: An analysis of variance test for normality (complete samples). Biometrika **52**, 591–611 (1965)

[Spi83]   Spiegelhalter, D.J.: Diagnostic tests of distributional shape. Biometrika **70**(2), 401–409 (1983)

[STET01]   Spivey, M.J., Tyler, M.J., Eberhard, K.M., Tanenhaus, M.K.: Linguistically mediated visual search. Psychol. Sci. **12**, 282–286 (2001)

[SC09]   Staudte, M., Crocker, M.W.: Visual attention in spoken human-robot interaction. In Proceedings of International Conference on Human-Robot Interaction, pp. 77–84 (2009)

[SYW97]   Stiefelhagen, R., Yang, J., Waibel, A.: Tracking eyes and monitoring eye gaze. In: Workshop on Perceptual User Interfaces (1997)

[SMSK08]   Sugano, Y., Matsushita, Y., Sato, Y., Koike, H.: An incremental learning method for unconstrained gaze estimation. In: ECCV (2008)

[SKI+07]   Sugiyama, O., Kanda, T., Imai, M., Ishiguro, H., Hagita, N.: Natural deictic communication with humanoid robots. In: Proceedings of International Conference on Intelligent Robots Systems (2007)

[TKA02]   Tan, K.H., Kriegman, D.J., Ahuja, N.: Appearance-based eye gaze estimation. In: IEEE Workshop on Applications of Computer Vision (2002)

[Tat07]   Tatler, B.W.: The central fixation bias in scene viewing: selecting an optimal viewing position independently of motor biases and image feature distributions. J. Vis. **7**(14) (2007)

[TB00]   Tjong Kim Sang, E.F., Buchholz, S.: Introduction to the CoNLL-2000 shared task: chunking. In: Proceedings of International Workshop on Computational Natural Language Learning, pp. 127–132 (2000)

[Tom03]   Tomasello, M.: Constructing a Language: A Usage-Based Theory of Language Acquisition. Harvard University Press (2003)

[TTDC06]   Triesch, J., Teuscher, C., Deák, G.O., Carlson, E.: Gaze following: why (not) learn it? Dev. Sci. **9**(2), 125–147 (2006)

[TCC+09]   Tseng, P.H., Carmi, R., Cameron, I.G.M., Munoz, D.P.: Itti, L.: Quantifying center bias of observers in free viewing of dynamic natural scenes. J. Vis. **9**(7) (2009)

[TCW+95]   Tsotsos, J.K., Culhane, S.M., Wai, W.Y.K., Lai, Y., Davis, N., Nuflo, F.: Modeling visual attention via selective tuning. Artif. Intell. **78**(1–2), 507–545 (1995)

[VSG12]   Valenti, R., Sebe, N., Gevers, T.: Combining head pose and eye location information for gaze estimation. IEEE Trans. Image Process. **21**, 802–815 (2012)

[vdWSV07]   van de Weijer, J., Schmid, C., Verbeek, J.J.: Learning color names from real-world images. In: Proceedings of International Conference on Computer Vision Pattern Recognition (2007)

[VT08]   Verbeek, J., Triggs, B.: Scene segmentation with crfs learned from partially labeled images. Proceedings of Annual Conference on Neural Inforrmation Processing Systems **20**, 1553–1560 (2008)

[VK89]   Vogel, R.M., Kroll, C.N.: Low-flow frequency analysis using probability-plot correlation coefficients. J. Water Resour. Plan. Manag. **115**(3), 338–357 (1989)

[VS08]   Voit, M., Stiefelhagen, R.: Deducing the visual focus of attention from head pose estimation in dynamic multi-view meeting scenarios. In: Proceedings of Internaional Conference on Multimodal Interfaces, pp. 173–180 (2008)

[WJ08]   Wainwright, M.J., Jordan, M.I.: Graphical Models, Exponential Families, and Variational Inference. Now Publishers Inc., Hanover (2008)

[Wel11]   Welke, K.: Memory-based active visual search for humanoid robots. Ph.D. Dissertation, Karlsruhe Institute of Technology (2011)

[WAD09]   Welke, K., Asfour, T., Dillmann, R.: Active multi-view object search on a humanoid head. In: Proceedings of International Conference on Robotic Automation (2009)

[Wol94]   Wolfe, J.M.: Guided search 2.0: a revised model of visual search. Psychon. Bull. Rev. **1**, 202–238 (1994)

[WCF89]   Wolfe, J.M., Cave, K., Franzel, S.: Guided search: an alternative to the feature integration model for visual search. J. Exp. Psychol.: Human Percept. Perform. **15**, 419–433 (1989)

[WHK+04] Wolfe, J.M., Horowitz, T.S., Kenner, N., Hyle, M., Vasan, N.: How fast can you change your mind? The speed of top-down guidance in visual search. Vis. Res. **44**, 1411–1426 (2004)

[YSS10] Yu, C., Scheutz, M., Schermerhorn, P.: Investigating multimodal real-time patterns of joint attention in an hri word learning task. In: Proceedings of International Conference on Human-Robot Interaction (2010)

[YScM+13] Yücel, Z., Salah, A.A., Meriçli, Ç., Meriçli, T., Valenti, R., Gevers, T.: Joint attention by gaze interpolation and saliency. IEEE Trans. Syst., Man, Cybern. **43**(3), 829–842 (2013)

Chapter 5
Conclusion

In addition to the discussion and presentation of future work in Sects. 3.6 and 4.5, let us briefly summarize our contributions and provide an outlook on ongoing and future work to conclude this book.

5.1 Summary

We derived several novel quaternion-based spectral visual saliency models (QDCT, ESR, ESW, and EPQFT), all of which perform state-of-the-art on three well-known eye tracking datasets. Furthermore, we proposed to decorrelate each image's color information as a preprocessing step for a wide variety of visual saliency models. We have shown that color space decorrelation can improve the performance by about 4 % (normalized) for eight visual saliency algorithms on three established datasets with respect to three complementing evaluation measures. Although an improvement of 4 % is far from drastic, it is nevertheless a considerable achievement, because we are not aware of any other method or preprocessing step that is able to consistently and significantly improve the performance of such a wide range of algorithms. Furthermore, we improved the state-of-the-art in predicting where people look when human faces are visible in the image. Compared to Cerf et al.'s approach, we were able to improve the performance by 8 % (i.e., 25.2 % normalized by the ideal AUC) with automatic face detections.

To realize auditory attention, we introduced a novel auditory saliency model that is based on the Bayesian surprise of each frequency. To allow for real-time computation on a robotic platform, we derived Gaussian surprise, which is efficient to calculate due to its simple closed form solution. Since we addressed a novel problem domain, we had to introduce a novel quantitative, application-oriented evaluation methodology and evaluated our model's ability to detect arbitrary salient auditory events. Our results show that Bayesian surprise can efficiently and reliably detect salient acoustic events, which is shown by F_1, F_2, and F_4 scores of 0.767, 0.892, and 0.967.

© Springer International Publishing Switzerland 2016
B. Schauerte, *Multimodal Computational Attention for Scene Understanding and Robotics*, Cognitive Systems Monographs 30,
DOI 10.1007/978-3-319-33796-8_5

We combined auditory and visual saliency in a biologically-plausible model based on crossmodal proto-objects to implement overt attention on a humanoid robot's head. We performed a series of behavioral experiments, which showed that our model exhibits the desired behaviors. Based on a formalization as multiobjective optimization problem, we introduced ego motion as a further criterion to plan which proto-object to attend next. This way, we were able to substantially reduce the amount of head ego motion while still preferring to attend the most salient proto-objects first. Our solution exhibits a low normalized cumulated joint angle distance (NCJAD) of 15.0 %, which represents that the chosen exploration order requires a low amount of ego motion to attend all proto-objects, and a high normalized cumulated saliency (NCS) of 83.3 %, which indicates that highly salient proto-objects are attended early.

We investigated how the spatial distribution of objects in images influences salient object detection. Here, we provided the first empirical justification for a Gaussian center bias. This is shown by a probability plot correlation coefficient (PPCC) of 0.9988 between a uniform distribution and the angular distribution of salient objects around the image center, and a PPCC of 0.9987 between a half-Gaussian distribution and the distribution of distances of salient objects to the image center. Then, we demonstrated that the performance of salient object detection algorithms can be substantially influenced by undocumented spatial biases. We debiased the region contrast algorithm and subsequently integrated a well-modeled Gaussian bias. This way, we achieved two goals: First, through integration of our explicit Gaussian bias, we improved the state-of-the-art in salient object detection for web images and at the same time quantified the influence of the center bias. Second, we derived the currently best unbiased salient object detection algorithm, which is advantageous for other application domains such as, e.g., surveillance and robotics.

We presented saliency models that are able to integrate multimodal signals such as pointing and spoken object descriptions to guide the attention in human-robot interaction. We started with an initial heuristic model that combines our spectral saliency detection with a probabilistic corridor of attention, i.e. the "probabilistic pointing cone", to reflect the spatial information given by pointing references. Additionally, we discussed a biologically-inspired neuron-based saliency model that is able to integrate knowledge about the target object's appearance into visual search. We outperform both models by training conditional random fields that integrate features such as, most importantly, our locally debiased region contrast, multi-scale spectral visual saliency with decorrelated color space, the probabilistic pointing cone, and target color models. This way, we are able to focus the correct target object in the initial focus of attention for 92.45 % of the images in the PointAT dataset, which does not provide spoken target descriptions, and 75.21 % for the ReferAT dataset, which includes spoken target references. This translates to an improvement of $+10.37$ % and $+25.21$ % compared to the heuristic and neuron-based saliency models, respectively.

Finally, we learn to determine objects or object parts that are being looked-at by persons in web images. This can be interpreted as a form of gaze following in web images. For this purpose, we integrated our work on salient object detection in web images and the interpretation of attentional signals in human-robot interaction. Consequently, we transferred our methods and train conditional random fields to integrate features such as, most importantly, spectral visual saliency, region contrast saliency, and a probabilistic corridor of interest that represents the observed gaze direction. This way, the looked-at target object is focused in the initial focus of attention for 66.17 % of images in a dataset that we collected from Flickr.

To quantify the performance of our approaches, we had to collect several datasets and even propose novel evaluation procedures, because we often addressed novel tasks, problems, and domains. We derived novel evaluation procedures for these tasks: First, we quantified the ability of our auditory saliency model to determine arbitrary salient acoustic events. Therefore, we relied on measures that are commonly used to evaluate salient object detection algorithms. Second, we introduced several novel evaluation measures to evaluate tradeoffs made by our multiobjective exploration path strategies. Furthermore, we proposed several measures to quantify the ability of saliency models to highlight target objects and focus the objects after a minimum amount of focus of attention shifts. We collected novel datasets for the following tasks: First, we created a dataset that consists of 60 videos to evaluate multiobjective exploration strategies. Second and third, we recorded two new datasets to evaluate how well we are able to guide our saliency model in human-robot interaction; in the absence (PointAT) and presence (ReferAT) of spoken target object information. For this purpose, fourth, we also gathered the Google-512 dataset to train our color term models. Fifth, to evaluate the identification and segmentation of gazed-at objects in web images, we collected the Gaze@Flickr dataset that we selected out of one million Flickr images.

5.2 Future Work

There are many aspects of high-level influences on visual saliency, search, and attention that represent interesting research directions such as, for example: How does coarse contextual information about the scene prime visual attention mechanisms and influence eye gaze patterns? Or, how do depicted pointing gestures and gaze directions in images influence gaze patterns of human observers? Here, it would be very interesting to compare the predictions of our top-down guided saliency models against human gaze behavior.

Furthermore, we have seen that human gaze estimates seem to depend on the visible image content. Accordingly, it seems that a very interesting research direction is to fuse gaze estimation and the detection of potential looked-at regions. In our opinion, an integrated approach should be able to jointly improve both estimates, i.e.. the estimated gaze direction and the predicted looked-at image region. Naturally, the same assumption and approach could also benefit pointing gestures and other directed non-verbal signals such as, e.g., head nods.

In our opinion, the currently most important open question for auditory and also audio-visual saliency models is a quantitative evaluation methodology based on human behavior. For example, it would be interesting to investigate the potential use of eye pupil dilation to evaluate bottom-up auditory saliency models. Furthermore, eye tracking experiments and datasets to investigate audio-visual saliency models and integration are still lacking. Consequently, the collection of a public benchmark dataset for audio-visual saliency models would represent a very valuable contribution to the field that could accelerate future research.

Our crossmodal proto-object model provides us with great flexibility to implement overt attention and saliency-based exploration mechanisms. However, since our hardware platform was stationary, we were only able to plan and evaluate head motion. Accordingly, one major open task is the implementation and evaluation of a parametric proto-object model as a basis for exploration mechanisms of non-stationary platforms. However, we are very confident that our model will prove to be viable in that scenario, the biggest challenge being sufficiently accurate robot self-localization that is necessary to update the proto-objects in the world model.

Appendix A
Applications

In the following, we briefly present further applications that use saliency models that are presented in the main body of this book. First, described in Sect. A.1, Martinez uses the Gaussian surprise model, see Sect. 3.3.1.2, to detect patient agitation in intensive care units. Second, described in Sect. A.2, Rybok relies on the quaternion image signature saliency model, see Sect. 3.2.1, and a notion of visual proto-objects, see Sect. 3.4, to improve the accuracy of activity recognition.

A.1 Patient Agitation

Appropriate patient sedation is a complex problem in intensive care units ICUs, because excessive sedation can threaten the patient's life while insufficient sedation can lead to excessive patient anxiety and agitation. The appropriate sedation protocol varies between patient and it does not just depend on easily measurable vital signs (e.g., heart rate), but also on behavioral cues that indicate signs for agitation, which are usually recorded by the nursing staff. In order to automate and improve the incorporation of such behavioral cues, it has been suggested to use computer vision systems to continuously monitor the patient's body and face for signs of stress, discomfort, or abnormalities. Here, the fact that such an automated system provides quantified and more objective measurements is a welcomed side effect. The most common behavioral cues are patient agitation patterns, because they are meaningful, robust to occlusions, and relatively easy to measure. For this purpose, Martinez proposed to apply surprise to detect and quantify agitation patterns that become apparent in a patient's face.

© Springer International Publishing Switzerland 2016
B. Schauerte, *Multimodal Computational Attention for Scene Understanding and Robotics*, Cognitive Systems Monographs 30,
DOI 10.1007/978-3-319-33796-8

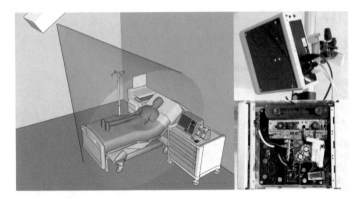

Fig. A.1 An illustration of Martinez's medical recording device (MRD) that is used to monitor patients in ICUs. The device uses stereo and depth cameras to record the entire body, while an additional high-resolution camera focuses on the face region. Image from and used with permission from Manel Martinez

A.1.1 Method

The first step in Martinez's framework is to determine the bed position, which is assumed to be roughly centered in the sensor setups field of view, see Fig. A.1. Then, the bed plane is estimated by a segmentation via region growing based on the depth map. Then two features are extracted from each image frame: First, the depth camera's information is used to calculate the bed occupancy feature, which measures the occupied volume over the bed plane. This feature is suited to detect events such as when the patient enters and exits the bed. Furthermore, it can also be used as a feature for body agitation and—given sufficient accuracy of the depth sensor—breath patterns. Second, the face camera is used to calculate a measure for agitation signals that are visible in the patient face. For this purpose, each image is resized to 32×32 px and Gaussian surprise is calculated for each pixel with a history length of 25 frames which is equivalent to 500 ms. This feature is suited to detect facial agitation patterns that are evident when the patient shows signs of discomfort.

A.1.2 Qualitative Evaluation

Due to the subjective behavior of the measurements and the lack of a public database or even a common evaluation methodology, it is impossible to quantitatively compare Martinez's results to alternative approaches. Instead, a qualitative behavioral experimental evaluation was performed. For this purpose, Martinez enacted and simulated a series of scenarios and compared the observed system behavior with the desired

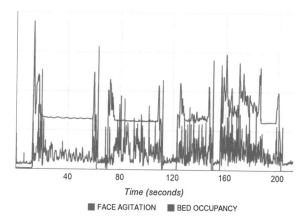

Fig. A.2 Surprise-based face agitation estimation (*blue*) in a simulation of 4 scenarios. From second 10 to 60: Sleeping relaxed shows an almost flat bed occupancy indicator and low agitation levels in the face. From second 70 to 110: Sleeping with pain expressions is not reflected in the volumetric information, but it is detected by the face agitation levels. From second 120 to 145: Being restless in bed is reflected by a clear response in both indicators. From second 145 to 200: Strong compulsions ending with an accident and sudden loss of consciousness. Image from and used with permission from Manel Martinez

behavior, see Fig. A.2. It was shown that the system provides reasonable behavioral descriptions of scenarios. In contrast to prior art [BHCS07, GCAS+04], the system is able to achieve this without relying special markers, invasive measures, or the need to control the illumination conditions.

A.2   Activity Recognition

Action and activity recognition is an important computer vision task with many potential application areas such as, for example, human-robot interaction, surveillance and multimedia retrieval. It is important to understand what differentiates the concepts of actions and activities. While the first describe simple motion events (e.g., "person stands up"), the latter describe complex action sequences (e.g., "person cleans kitchen") that form an activity. According to action identification theory, actions and, as a consequence, activities are not just defined by motion patterns but derive their meaning from context [VW87]. For example, the motion patterns for "wiping" and "waving" can look very similar and hard to distinguish without the context in which they are performed. Consequently, it can be necessary to incorporate—among other contextual cues—the location where an action is performed or which objects are manipulated in the activity classification process.

Most work on activity recognition does not integrate contextual knowledge or requires specifically trained detectors. However, such detectors require a considerable amount of manually annotated training data, which is costly to acquire and makes it hard to transfer the activity recognition systems to new application areas and domains. As an alternative, Rybok proposes to use salient proto-objects to detect candidate objects, object parts, or groups of objects (see Sect. 3.4) that are potentially relevant for the activity. This approach makes it possible to integrate contextual object knowledge into the activity recognition based on unsupervised methods, i.e. without the need for accurate object labels or detectors.

A.2.1 Method

Rybok relies on the QDCT image signatures, see Sect. 3.2.1, to calculate the visual saliency of each frame in a video sequence, see Fig. A.3. To determine the image's proto-object regions, Felzenszwalb's graph-based algorithm [FH04] is used to segment each frame. Following the classical winner-take-all method for attentional shifts and inhibition of return (see Sect. 2.1.1), Rybok iteratively extracts the image segment that contains the most salient peak and then inhibits the saliency at the segment's location. This is repeated, until the saliency map's maximum saliency value either falls below 70 % of its initial maximum or the 30 most salient segments have been extracted. The extracted segments form the set of proto-object regions that serves as context (i.e., as object or object part candidates) for the activity recognition. To this end, the appearance of each extracted proto-object region is encoded by Dalal and Trigg's histogram of oriented gradients (HOG) [DT05]. Given the activity recognition training sequences, the proto-object HOG feature vectors are clustered with k-means to generate a proto-object codebook.

The classification of motion patterns is based on Laptev et al.'s space time interest points (STIP) [LMSR08]. Laptev et al.'s Harris 3D interest point detection is used to determine interesting points in space and time. Either the histogram of optical flow (HOF) alone or a combination of HOF and HOG is used as feature vector to describe each interest point. The HOG descriptor in this context is different from Dalal and Trigs's HOG descriptor [DT05], because it accumulates the gradients within the spatio-temporal STIP region.

(a)   **(b)**   **(c)**   **(d)**

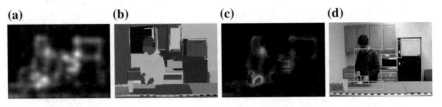

Fig. A.3 Example of the proto-object extraction approach. Image from [RSAHS14]. **a** Saliency map. **b** Image segmentation. **c** Saliency weighted segments. **d** Proto-object regions

Each image sequence is represented by a bag-of-words feature vector with a 1000-element codebook for motion features (HOG or HOG-HOF) and a 200-element codebook for proto-objects. Given these features, a linear SVM is trained to classify video sequences. To boost the performance, the feature vector is normalized and then each element is raised to the power of $\alpha = 0.3$ (cf. [RR13]).

Table A.1 Activity recognition results on the (a) URADL, (b) CAD-120, and (c) KIT datasets

(a) URADL

Method	Accuracy (%)
HOF	79.3
HOF and all segments	86.7
HOF and proto-objects	97.7
HOGHOF	94.0
HOGHOF and all segments	94.7
HOGHOF and proto-objects	**100.0**
Matikainen et al. [MHS10], 2010	70.0
Messing et al. [MPK09], 2009	89.0
Prest et al. [PFS12], 2012	92.0
Wang et al. [WCW11], 2011	96.0
Yi and Lin [YL13], 2013	98.0

(b) CAD-120

HOF	72.6
HOF and all segments	75.0
HOF and proto-objects	**79.0**
HOGHOF	70.0
HOGHOF and all segments	72.6
HOGHOF and proto-objects	77.4
Sung et al. [SPSS12], 2012	26.4
Koppula et al. [KS13], 2013	75.0

(c) KIT

HOF	85.6
HOF and proto-objects	**88.7**
HOGHOF	86.6
HOGHOF and proto-objects	88.5
Rybok et al. [RFHS11], 2011	84.9
Onofri et al. [OSI13], 2013	88.3

As can be seen, the combination of contextual proto-object information and simple HOG and HOG-HOF features provides state-of-the-art performance

A.2.2 Results

Rybok evaluates the approach on three activity recognition benchmark datasets: URADL [MPK09], CAD-120 [KS13], and KIT Robo-Kitchen [RFHS11]. As can be seen in Table A.1, the saliency-driven approach is able improve the state-of-the-art on all three datasets, although the employed motion features are relatively common and simple. To demonstrate the benefit of saliency-driven object candidate extraction, Rybok compares to an alternative approach in which all image segments are used as contextual information ("all segments", Table A.1). Although the performance achieved with all image segments is better than the model without context information, it is clear that the saliency-driven image region selection provides a substantial performance benefit, see Table A.1.

Appendix B
Dataset Overview

Throughout this book, we rely on several datasets to evaluate our algorithms. In the following, we provide a short overview of these datasets.

Main datasets (Chap. 3):

1. Bruce/Toronto: Eye tracking (Sect. 3.2.1.6)
   This dataset [BT09] contains 120 color images depicting indoor and outdoor scenes. The dataset contains eye tracking data of 20 subjects (4 s, free-viewing).
2. Judd/MIT: Eye tracking (Sect. 3.2.1.6)
   This dataset contains 1003 images of varying resolutions [JEDT09] that were collected from Flickr and the LabelMe database. Eye tracking data was recorded for 15 subjects (3 s, free-viewing).
3. Kootstra: Eye tracking (Sect. 3.2.1.6)
   This dataset [KNd08] contains 100 images that were collected from the McGill calibrated color image database [OK04]. The images were shown to 31 subjects (free-viewing).
4. Cerf/FIFA: Eye tracking (Sect. 3.2.3.3)
   To evaluate the influence of faces on human visual attention, this dataset [CFK09] consists of eye tracking data (2 s, free-viewing) of 9 subjects for 200 images of which 157 contain one or more faces.
5. CLEAR2007: Acoustic events (Sect. 3.3.2.2)
   The CLEAR2007 acoustic event detection dataset [CLE, TMZ+07]) contains recordings of meetings in a smart room. A human analyst marked and classified occurring acoustic events that were remarkable enough to "pop-out". 14 acoustic event classes were identified and tagged (e.g., "laughter", "door knocks", "phone ringing" and "key jingling"). Events that could not be identified by the human analyst were tagged as "unknown".

© Springer International Publishing Switzerland 2016
B. Schauerte, *Multimodal Computational Attention for Scene Understanding and Robotics*, Cognitive Systems Monographs 30,
DOI 10.1007/978-3-319-33796-8

Main datasets (Chap. 4):

6. IROS2012: Exploration strategies (Sect. 3.5)
   To evaluate scene exploration strategies [KSKS12], this dataset consists of 60 videos (30 s each), in which specific sequences were re-enacted in three scenarios: office scenes, breakfast scenes, and neutral scenes.

7. PointAT: Pointing (Sect. 4.3.1.4)
   This dataset contains 220 instances of 3 persons pointing at objects in an office environment and conference room [SRF10]. Pointed-at objects were predicted online while recording the dataset and used to automatically zoom on the target object, which additionally makes it possible to evaluate the influence of foveation on object recognition.

8. ReferAT: Pointing and language (Sect. 4.3.2.5)
   This dataset contains 242 multimodal referring acts (composed of pointing gestures and spoken object descriptions) that were performed by 5 persons referring to a set of 28 objects in a meeting room [SF10a]. The objects were chosen in such a way that, in most situations, object names and colors are the most discriminant verbal cues for referring-to the referent.

9. Gaze@Flickr: Gaze (Sect. 4.4.2)
   Our Gaze@Flickr dataset contains 863 Flickr images that contain 1221 gaze references, i.e. persons gazing at a target object. The dataset provides annotated head regions of the gazing persons as well as two different gaze directions that were annotated under different viewing conditions.

10. MSRA: Salient objects (Sect. 4.2.1)
    MSRA is the most widely used dataset to evaluate salient object detection. It has been created by Achanta et al. and Liu et al. [AHES09, LSZ+07] and consists of 1000 images with binary segmentation masks of the target object.

Additional datasets:

A. Google-512: Color terms (Sect. 4.3.2.3)
   We use our Google-512 dataset [SF10b] to learn color term models. The dataset consists of 512 images for each of the eleven basic English color terms. The images were collected using Google's image search. The learned color term models are commonly evaluated on another dataset: Weijer et al.'s e-Bay dataset [vdWSV07].

B. Brown: Language (Sect. 4.3.2.1)
   We use the Brown corpus [Fra79] and its annotated part-of-speech tags to train a Brill tagger [Bri95], which we use to determine noun-phrases and their constituents with a shallow parser which is based on regular expressions. The Brown corpus is a general text collection, i.e. general corpus, that contains 500 samples of English text, compiled from works that were published in the United States in 1961.

Additionally, we performed behavioral experiments to evaluate our audio-visual overt attention system, see Sect. 3.4. Furthermore, we collected some datasets that are not relevant to the content in this book such as, e.g., the Flower Box dataset for visual obstacle detection and avoidance [KSS13].

Appendix C
Color Space Decorrelation: Full Evaluation

In the following, we present further color space decorrelation evaluation results that complement our evaluation and discussion in Sect. 3.2.2.2. For this purpose, we provide the evaluation results of our baseline algorithms for three evaluation measures, see Table C.1. We provide results for the following color spaces: RGB, CIE Lab, CIE XYZ, ICOPP (e.g., [GMZ08]), LMS [SG31], and Gauss [GvdBSG01]. Furthermore, we use statistical tests (see Sect. 3.2.2.2) to test the performance of each algorithm on the original color space against the performance based on the decorrelated color space ("better", "better or equal", "probably equal", "equal or worse", and "worse").

Evaluation Measures

In this book, we focused on the AUC evaluation measure in the main document, since it is the most established measure. However, in our extended evaluation results for color space decorrelation (Appendix C), we use the CC and NSS as complementary evaluation measures to show that color space decorrelation is beneficial as quantified by all three evaluation measure classes [RDM+13], see Sect. 3.2.1.6. Let us present the three evaluation measures in more detail.

The shuffled, bias-correcting AUROC or shorter AUC measure (see, e.g., [HHK12]) tries to compensate for biases such as, e.g., the center-bias that is commonly found in eye tracking datasets. To this end, it defines a positive and a negative set of eye fixations for each image. The positive sample set contains the fixation points of all subjects on that image. The negative sample set contains the union of all eye fixation points across all other images from the same dataset. To calculate the AUROC, each saliency map is thresholded and the resulting binary map can be seen as being a binary classifier that tries to classify positive and negative samples. Sweeping over all thresholds leads to the ROC curves and defines the area under the ROC curve. When using the AUROC as a measure, the chance level is 0.5 (random classifier), values <0.5 indicate negative correlation, values >0.5 represent positive correlation, and a AUROC of 1 means perfect classification (Tables C.2, C.5 and C.8).

© Springer International Publishing Switzerland 2016
B. Schauerte, *Multimodal Computational Attention for Scene Understanding and Robotics*, Cognitive Systems Monographs 30,
DOI 10.1007/978-3-319-33796-8

Table C.1 Performance of selected baseline algorithms on the Kootstra, Judd/MIT, and Bruce/Toronto datasets

Method	Bruce/Toronto			Kootstra			Judd/MIT		
	ROC	CC	NSS	ROC	CC	NSS	ROC	CC	NSS
CAS'12	0.692	0.370	1.255	0.603	0.246	0.544	0.662	0.235	0.948
CCH'12	0.666	0.268	0.905	0.583	0.219	0.478	0.648	0.218	0.873
JEDA'09	0.624	0.420	1.379	0.549	0.307	0.651	0.665	0.342	1.351
AIM'09	0.666	0.261	0.898	0.574	0.176	0.383	0.637	0.184	0.747
GBVS'07	0.660	0.420	1.381	0.558	0.220	0.458	0.584	0.174	0.693
COH'06	0.650	0.310	0.990	0.547	0.263	0.510	0.697	0.210	0.990
IK'98	0.645	0.393	1.293	0.574	0.279	0.585	0.636	0.261	1.039
iNVT'98	0.544	0.155	0.553	0.518	0.092	0.210	0.536	0.099	0.418
Chance	0.5	$\rightarrow 0$	≤ 0	0.5	$\rightarrow 0$	≤ 0	0.5	$\rightarrow 0$	≤ 0

The linear correlation coefficient (CC) is a measure for the strength of a linear relationship between two variables. Let G denote the groundtruth saliency map that is generated by adding a Gaussian blur to the recorded eye fixations and S the algorithm's saliency map [JOvW+05, RBC06], then CC$(G, S) = \frac{cov(G,S)}{\sigma_G \sigma_S}$, where σ_G and σ_S are the standard deviations of G and S, respectively. A CC close to $+1$ or -1 indicates an almost perfectly linear relationship between the prediction S and groundtruth G. As the CC approaches 0 there is less of a relationship, i.e. it is closer to being uncorrelated (Tables C.3, C.6 and C.9).

The normalized scanpath saliency (NSS) is the average saliency at human eye fixations in an algorithm's saliency map. To make the values comparable, the saliency map is normalized to have zero mean and unit standard deviation [PIIK05, PLN02], i.e. a NSS of 1 means that the predicted saliency at recorded eye fixations is one standard deviation above average. Consequently, an NSS ≥ 1 indicates that the saliency map has significantly higher saliency values at locations that were fixated by the human subjects than at other locations. An NSS ≤ 0 means that the predicted saliency does not predict eye fixations better than picking random image locations, i.e. chance (Tables C.4, C.7, and C.10).

Table C.2 Color space decorrelation results as quantified by the AUC evaluation measure on the Bruce/Toronto dataset

AUC Method	RGB raw	PCA	ZCA	Lab raw	PCA	ZCA	ICOPP raw	PCA	ZCA
CCH'12	0.6661	**0.7031**	0.6974	0.6979	0.7061	**0.7072**	0.6881	**0.7032**	0.7019
QDCT	0.7033	**0.7157**	0.7149	0.7158	0.7187	**0.7210**	0.7135	0.7140	**0.7175**
EPQFT	0.7003	0.7142	**0.7158**	0.7154	0.7180	**0.7212**	0.7112	0.7118	**0.7156**
DCT'11	0.6915	**0.7196**	0.7121	0.7126	**0.7208**	0.7207	0.7114	**0.7184**	0.7166
AC'09	0.5406	0.5608	**0.5780**	0.5541	0.5609	**0.5735**	0.5510	0.5543	**0.5702**
GBVS'07	0.6030	**0.6620**	0.6614	0.6371	**0.6665**	0.6655	0.6374	**0.6637**	0.6617
PFT'07	0.6952	**0.7196**	0.7135	0.7141	**0.7226**	**0.7226**	0.7128	0.7179	**0.7189**
IK'98	0.6410	0.6723	**0.6772**	0.6612	0.6734	**0.6814**	0.6636	0.6721	**0.6756**

AUC Method	GAUSS raw	PCA	ZCA	XYZ raw	PCA	ZCA	CAT02LMS raw	PCA	ZCA
CCH'12	0.6959	0.7019	**0.7038**	0.6303	**0.6664**	0.6578	0.6306	**0.6657**	0.6563
QDCT	0.7135	0.7153	**0.7171**	0.6665	**0.6754**	0.6708	0.6657	**0.6749**	0.6704
EPQFT	0.7110	0.7130	**0.7168**	0.6642	**0.6746**	0.6688	0.6638	**0.6739**	0.6678
DCT'11	0.7097	0.7190	**0.7205**	0.6617	**0.6811**	0.6682	0.6612	**0.6804**	0.6667
AC'09	0.5576	0.5620	**0.5622**	0.5263	0.5433	**0.5474**	0.5280	0.5438	**0.5451**
GBVS'07	0.6417	0.6618	**0.6619**	0.5784	**0.6207**	0.6203	0.5794	0.6203	**0.6206**
PFT'07	0.7114	0.7180	**0.7200**	0.6619	**0.6820**	0.6701	0.6619	**0.6818**	0.6678
IK'98	0.6702	0.6719	**0.6738**	0.6110	0.6393	**0.6474**	0.6099	0.6389	**0.6472**

Please refer to Table 3.3 for a color legend

Table C.3 Color space decorrelation results as quantified by the CC evaluation measure on the Bruce/Toronto dataset

CC Method	RGB raw	PCA	ZCA	Lab raw	PCA	ZCA	ICOPP raw	PCA	ZCA
CCH'12	0.2688	**0.3490**	0.3360	0.3408	0.3522	**0.3543**	0.3334	**0.3507**	0.3491
QDCT	0.3484	0.3943	**0.4034**	0.3988	0.4017	**0.4087**	0.3909	0.3933	**0.3991**
EPQFT	0.3023	0.3574	**0.3769**	0.3638	0.3627	**0.3786**	0.3447	0.3501	**0.3575**
DCT'11	0.3200	**0.4213**	0.4019	0.4068	**0.4220**	0.4213	0.4063	**0.4191**	0.4168
AC'09	0.0485	0.0861	**0.1091**	0.0764	0.0929	**0.1151**	0.0629	0.0716	**0.1047**
GBVS'07	0.2385	0.3364	**0.3391**	0.2820	**0.3462**	0.3432	0.2890	**0.3419**	0.3383
PFT'07	0.2905	**0.4007**	0.3840	0.3869	0.4046	**0.4050**	0.3857	**0.3998**	**0.3998**
IK'98	0.3204	0.3865	**0.3938**	0.3557	0.3848	**0.3965**	0.3640	**0.3877**	0.3851

CC Method	GAUSS raw	PCA	ZCA	XYZ raw	PCA	ZCA	CAT02LMS raw	PCA	ZCA
CCH'12	0.3305	**0.3490**	0.3465	0.1836	**0.2620**	0.2485	0.1847	**0.2601**	0.2458
QDCT	0.3908	0.3918	**0.3949**	0.2637	0.2967	**0.3087**	0.2671	0.2962	**0.3102**
EPQFT	0.3516	0.3507	**0.3607**	0.2241	0.2606	**0.2759**	0.2273	0.2603	**0.2774**
DCT'11	0.3999	**0.4192**	0.4159	0.2517	**0.3359**	0.3119	0.2546	**0.3347**	0.3107
AC'09	0.0813	**0.0917**	0.0862	0.0160	0.0285	**0.0543**	0.0185	0.0295	**0.0498**
GBVS'07	0.3065	**0.3360**	0.3358	0.1760	0.2669	**0.2674**	0.1785	**0.2673**	0.2671
PFT'07	0.3776	**0.3974**	0.3963	0.2190	**0.3116**	0.2883	0.2217	**0.3102**	0.2870
IK'98	0.3813	0.3835	**0.3852**	0.2719	0.3257	**0.3458**	0.2738	0.3250	**0.3459**

Please refer to Table 3.3 for a color legend

Table C.4 Color space decorrelation results as quantified by the NSS evaluation measure on the Bruce/Toronto dataset

NSS Method	RGB			Lab			ICOPP		
	raw	PCA	ZCA	raw	PCA	ZCA	raw	PCA	ZCA
CCH'12	0.9052	**1.1700**	1.1268	1.1415	1.1801	**1.1880**	1.1148	**1.1730**	1.1705
QDCT	1.1860	1.3376	**1.3665**	1.3558	1.3636	**1.3884**	1.3271	1.3347	**1.3533**
EPQFT	1.0342	1.2187	**1.2804**	1.2429	1.2349	**1.2917**	1.1752	1.1919	**1.2173**
DCT'11	1.0905	**1.4304**	1.3578	1.3861	**1.4357**	1.4315	1.3782	**1.4246**	1.4114
AC'09	0.1822	0.3203	**0.3950**	0.2807	0.3397	**0.4173**	0.2343	0.2672	**0.3864**
GBVS'07	0.7892	1.1080	**1.1185**	0.9273	**1.1412**	1.1309	0.9463	**1.1268**	1.1147
PFT'07	0.9946	**1.3664**	1.3052	1.3248	**1.3822**	1.3821	1.3143	**1.3651**	1.3604
IK'98	1.0651	1.2779	**1.3000**	1.1771	1.2729	**1.3079**	1.2055	**1.2816**	1.2714

NSS Method	GAUSS			XYZ			CAT02LMS		
	raw	PCA	ZCA	raw	PCA	ZCA	raw	PCA	ZCA
CCH'12	1.1093	**1.1718**	1.1631	0.6226	**0.8806**	0.8325	0.6260	**0.8737**	0.8223
QDCT	1.3285	1.3293	**1.3399**	0.9051	1.0099	**1.0413**	0.9160	1.0079	**1.0468**
EPQFT	1.2014	1.1941	**1.2288**	0.7752	0.8940	**0.9348**	0.7860	0.8933	**0.9399**
DCT'11	1.3583	**1.4253**	1.4127	0.8651	**1.1426**	1.0509	0.8747	**1.1376**	1.0457
AC'09	0.2998	**0.3397**	0.3176	0.0825	0.1258	**0.2038**	0.0898	0.1289	**0.1892**
GBVS'07	1.0093	**1.1071**	1.1062	0.5929	0.8785	**0.8814**	0.6008	0.8801	**0.8805**
PFT'07	1.2895	**1.3560**	1.3516	0.7583	**1.0637**	0.9772	0.7679	**1.0587**	0.9717
IK'98	1.2660	1.2694	**1.2722**	0.9132	1.0772	**1.1421**	0.9191	1.0748	**1.1423**

Please refer to Table 3.3 for a color legend

Table C.5 Color space decorrelation results as quantified by the AUC evaluation measure on the Judd/MIT dataset

AUC Method	RGB			Lab			ICOPP		
	raw	PCA	ZCA	raw	PCA	ZCA	raw	PCA	ZCA
CCH'12	0.6480	0.6696	**0.6708**	0.6674	**0.6733**	0.6722	0.6595	**0.6705**	0.6702
QDCT	0.6517	0.6608	**0.6625**	0.6599	0.6610	**0.6625**	0.6585	0.6593	**0.6613**
EPQFT	0.6484	0.6590	**0.6621**	0.6579	0.6581	**0.6609**	0.6547	0.6558	**0.6583**
DCT'11	0.6440	**0.6641**	0.6608	0.6581	**0.6645**	0.6638	0.6577	**0.6632**	0.6627
AC'09	0.5306	0.5513	**0.5810**	0.5493	0.5514	**0.5592**	0.5452	0.5492	**0.5585**
GBVS'07	0.5846	**0.6343**	0.6327	0.6207	**0.6367**	0.6362	0.6162	**0.6349**	0.6342
PFT'07	0.6449	**0.6652**	0.6627	0.6597	**0.6653**	0.6650	0.6590	0.6639	**0.6647**
IK'98	0.6367	0.6572	**0.6585**	0.6508	0.6581	**0.6582**	0.6493	0.6556	**0.6564**

AUC Method	GAUSS			XYZ			CAT02LMS		
	raw	PCA	ZCA	raw	PCA	ZCA	raw	PCA	ZCA
CCH'12	0.6657	0.6700	**0.6705**	0.6207	**0.6460**	0.6457	0.6202	0.6452	**0.6458**
QDCT	0.6573	0.6600	**0.6610**	0.6262	0.6385	**0.6448**	0.6262	0.6380	**0.6455**
EPQFT	0.6552	0.6570	**0.6595**	0.6240	0.6362	**0.6434**	0.6240	0.6353	**0.6446**
DCT'11	0.6549	**0.6639**	0.6632	0.6216	**0.6452**	0.6430	0.6218	**0.6450**	0.6429
AC'09	0.5459	0.5526	**0.5532**	0.5205	0.5371	**0.5602**	0.5214	0.5372	**0.5587**
GBVS'07	0.6134	**0.6343**	**0.6343**	0.5671	**0.6124**	0.6101	0.5670	**0.6125**	0.6103
PFT'07	0.6568	**0.6650**	0.6649	0.6225	**0.6461**	0.6440	0.6224	**0.6456**	0.6439
IK'98	0.6480	0.6565	**0.6570**	0.6134	0.6372	**0.6407**	0.6131	0.6365	**0.6409**

Please refer to Table 3.3 for a color legend

Table C.6 Color space decorrelation results as quantified by the CC evaluation measure on the Judd/MIT dataset

CC Method	RGB raw	PCA	ZCA	Lab raw	PCA	ZCA	ICOPP raw	PCA	ZCA
CCH'12	0.2180	0.2363	**0.2389**	0.2317	**0.2404**	0.2343	0.2233	**0.2390**	0.2335
QDCT	0.2206	**0.2350**	0.2332	0.2345	**0.2351**	0.2349	0.2328	0.2338	**0.2340**
EPQFT	0.1918	0.2105	**0.2124**	0.2101	0.2097	**0.2114**	0.2050	0.2071	**0.2075**
DCT'11	0.2084	**0.2394**	0.2313	0.2336	**0.2389**	0.2372	0.2339	**0.2391**	0.2368
AC'09	0.0291	0.0494	**0.0873**	0.0490	0.0504	**0.0592**	0.0410	0.0469	**0.0566**
GBVS'07	0.1743	**0.2372**	0.2360	0.2203	**0.2393**	0.2389	0.2135	**0.2393**	0.2384
PFT'07	0.1860	**0.2195**	0.2132	0.2137	**0.2188**	0.2176	0.2135	**0.2193**	0.2181
IK'98	0.2616	0.2907	**0.2908**	0.2841	**0.2916**	0.2910	0.2832	**0.2903**	0.2885

CC Method	GAUSS raw	PCA	ZCA	XYZ raw	PCA	ZCA	CAT02LMS raw	PCA	ZCA
CCH'12	0.2365	**0.2372**	0.2360	0.1785	0.2091	**0.2101**	0.1785	0.2079	**0.2101**
QDCT	0.2317	**0.2344**	**0.2344**	0.1822	0.2015	**0.2100**	0.1830	0.2009	**0.2130**
EPQFT	0.2070	0.2086	**0.2107**	0.1586	0.1789	**0.1915**	0.1593	0.1781	**0.1956**
DCT'11	0.2310	**0.2391**	0.2384	0.1753	**0.2152**	0.2107	0.1761	**0.2150**	0.2110
AC'09	0.0421	0.0500	**0.0514**	0.0203	0.0304	**0.0644**	0.0208	0.0299	**0.0664**
GBVS'07	0.2102	0.2375	**0.2378**	0.1393	**0.2108**	0.2063	0.1395	**0.2104**	0.2064
PFT'07	0.2109	**0.2196**	0.2193	0.1557	**0.1984**	0.1950	0.1564	**0.1978**	0.1952
IK'98	0.2825	0.2901	**0.2909**	0.2259	0.2626	**0.2694**	0.2266	0.2619	**0.2700**

Please refer to Table 3.3 for a color legend

Table C.7 Color space decorrelation results as quantified by the NSS evaluation measure on the Judd/MIT dataset

NSS Method	RGB raw	PCA	ZCA	Lab raw	PCA	ZCA	ICOPP raw	PCA	ZCA
CCH'12	0.8736	0.9543	**0.9661**	0.9357	**0.9707**	0.9464	0.9033	**0.9648**	0.9438
QDCT	0.8895	**0.9469**	0.9392	0.9444	**0.9472**	0.9464	0.9380	0.9426	**0.9432**
EPQFT	0.7780	0.8535	**0.8599**	0.8506	0.8499	**0.8565**	0.8308	0.8401	**0.8409**
DCT'11	0.8397	**0.9632**	0.9304	0.9395	**0.9611**	0.9539	0.9413	**0.9625**	0.9530
AC'09	0.1259	0.2130	**0.3703**	0.2099	0.2167	**0.2528**	0.1792	0.2031	**0.2418**
GBVS'07	0.6939	**0.9460**	0.9410	0.8769	**0.9538**	0.9526	0.8517	**0.9545**	0.9511
PFT'07	0.7539	**0.8882**	0.8624	0.8642	**0.8852**	0.8801	0.8649	**0.8875**	0.8830
IK'98	1.0391	1.1531	**1.1540**	1.1260	**1.1573**	1.1544	1.1236	**1.1523**	1.1450

NSS Method	GAUSS raw	PCA	ZCA	XYZ raw	PCA	ZCA	CAT02LMS raw	PCA	ZCA
CCH'12	0.9533	**0.9575**	0.9532	0.7153	0.8457	**0.8516**	0.7148	0.8408	**0.8516**
QDCT	0.9330	**0.9445**	0.9443	0.7348	0.8147	**0.8509**	0.7380	0.8122	**0.8635**
EPQFT	0.8383	0.8452	**0.8539**	0.6432	0.7273	**0.7800**	0.6461	0.7238	**0.7975**
DCT'11	0.9293	**0.9623**	0.9588	0.7060	**0.8705**	0.8531	0.7089	**0.8694**	0.8543
AC'09	0.1820	0.2150	**0.2214**	0.0881	0.1320	**0.2785**	0.0899	0.1298	**0.2862**
GBVS'07	0.8374	0.9470	**0.9484**	0.5549	**0.8440**	0.8256	0.5555	**0.8425**	0.8258
PFT'07	0.8538	**0.8884**	0.8872	0.6312	**0.8075**	0.7937	0.6336	**0.8049**	0.7951
IK'98	1.1211	1.1509	**1.1543**	0.8969	1.0446	**1.0731**	0.8994	1.0419	**1.0756**

Please refer to Table 3.3 for a color legend

Table C.8 Color space decorrelation results as quantified by the AUC evaluation measure on the Kootstra dataset

AUC Method	RGB raw	PCA	ZCA	Lab raw	PCA	ZCA	ICOPP raw	PCA	ZCA
CCH'12	0.5838	0.6030	**0.6045**	0.6018	**0.6043**	0.6037	0.6027	0.6040	**0.6042**
QDCT	0.5974	0.6068	**0.6148**	0.6041	0.6049	**0.6088**	0.6045	0.6069	**0.6092**
EPQFT	0.5955	0.6050	**0.6140**	0.6021	0.6032	**0.6069**	0.6016	0.6050	**0.6070**
DCT'11	0.5891	**0.6148**	0.6143	0.6063	0.6126	**0.6147**	0.6074	0.6134	**0.6173**
AC'09	0.5415	0.5509	**0.5633**	0.5464	0.5487	**0.5544**	0.5463	0.5488	**0.5534**
GBVS'07	0.5584	**0.5897**	0.5879	0.5788	**0.5914**	0.5906	0.5764	**0.5912**	0.5901
PFT'07	0.5936	**0.6180**	0.6147	0.6087	0.6157	**0.6172**	0.6100	0.6159	**0.6190**
IK'98	0.5740	0.5951	**0.5965**	0.5882	0.5936	**0.5950**	0.5881	0.5943	**0.5966**

AUC Method	GAUSS raw	PCA	ZCA	XYZ raw	PCA	ZCA	CAT02LMS raw	PCA	ZCA
CCH'12	0.6025	0.6037	**0.6041**	0.5684	0.5834	**0.5862**	0.5682	0.5832	**0.5861**
QDCT	0.6050	0.6075	**0.6091**	0.5821	0.5960	**0.6047**	0.5829	0.5958	**0.6057**
EPQFT	0.6037	0.6064	**0.6068**	0.5820	0.5927	**0.6040**	0.5827	0.5928	**0.6045**
DCT'11	0.6058	**0.6161**	0.6155	0.5786	**0.6063**	0.6050	0.5793	**0.6069**	0.6065
AC'09	0.5484	0.5515	**0.5523**	0.5343	0.5436	**0.5522**	0.5340	0.5424	**0.5541**
GBVS'07	0.5736	0.5893	**0.5898**	0.5540	**0.5843**	0.5804	0.5539	**0.5846**	0.5805
PFT'07	0.6092	**0.6185**	0.6172	0.5816	**0.6092**	0.6072	0.5826	**0.6095**	0.6068
IK'98	0.5886	0.5957	**0.5959**	0.5643	0.5864	**0.5893**	0.5645	0.5867	**0.5890**

Please refer to Table 3.3 for a color legend

Table C.9 Color space decorrelation results as quantified by the CC evaluation measure on the Kootstra dataset

CC Method	RGB raw	PCA	ZCA	Lab raw	PCA	ZCA	ICOPP raw	PCA	ZCA
CCH'12	0.2193	**0.2204**	0.2194	0.2195	**0.2307**	0.2147	0.2130	**0.2197**	0.2120
QDCT	0.2449	0.2636	**0.2714**	0.2565	0.2599	**0.2664**	0.2574	0.2633	**0.2682**
EPQFT	0.2008	0.2209	**0.2333**	0.2138	0.2167	**0.2248**	0.2113	0.2200	**0.2248**
DCT'11	0.2300	**0.2766**	0.2752	0.2553	0.2722	**0.2760**	0.2598	0.2766	**0.2796**
AC'09	0.0610	0.0704	**0.0786**	0.0654	0.0681	**0.0724**	0.0609	**0.0703**	0.0694
GBVS'07	0.2201	**0.2844**	0.2798	0.2622	**0.2852**	0.2852	0.2590	**0.2868**	0.2847
PFT'07	0.1965	**0.2427**	0.2401	0.2203	0.2365	**0.2402**	0.2225	0.2412	**0.2445**
IK'98	0.2790	0.3136	**0.3203**	0.3012	0.3141	**0.3172**	0.2998	0.3130	**0.3179**

CC Method	GAUSS raw	PCA	ZCA	XYZ raw	PCA	ZCA	CAT02LMS raw	PCA	ZCA
CCH'12	**0.2195**	0.2175	0.2146	0.1800	0.1880	**0.1895**	0.1789	0.1879	**0.1911**
QDCT	0.2614	0.2665	**0.2679**	0.2088	0.2310	**0.2445**	0.2106	0.2308	**0.2466**
EPQFT	0.2207	**0.2254**	0.2246	0.1712	0.1917	**0.2124**	0.1726	0.1918	**0.2144**
DCT'11	0.2601	**0.2787**	0.2774	0.2035	**0.2518**	0.2513	0.2045	**0.2521**	0.2513
AC'09	0.0699	0.0689	**0.0704**	0.0574	0.0577	**0.0713**	0.0559	0.0571	**0.0722**
GBVS'07	0.2489	0.2830	**0.2853**	0.1922	**0.2609**	0.2533	0.1922	**0.2619**	0.2517
PFT'07	0.2270	**0.2436**	0.2418	0.1697	**0.2238**	0.2218	0.1711	**0.2242**	0.2217
IK'98	0.3050	0.3158	**0.3162**	0.2515	0.2907	**0.2960**	0.2535	0.2906	**0.2952**

Please refer to Table 3.3 for a color legend

Table C.10 Color space decorrelation results as quantified by the NSS evaluation measure on the Kootstra dataset

NSS Method	RGB raw	PCA	ZCA	Lab raw	PCA	ZCA	ICOPP raw	PCA	ZCA
CCH'12	0.4783	**0.4996**	0.4991	0.4972	**0.5219**	0.4863	0.4851	**0.4993**	0.4836
QDCT	0.5268	0.5760	**0.6042**	0.5588	0.5666	**0.5846**	0.5592	0.5771	**0.5911**
EPQFT	0.4384	0.4895	**0.5282**	0.4737	0.4801	**0.5010**	0.4644	0.4887	**0.5024**
DCT'11	0.4909	**0.6153**	0.6132	0.5562	0.6027	**0.6116**	0.5671	0.6156	**0.6221**
AC'09	0.1390	0.1650	**0.2014**	0.1561	0.1629	**0.1726**	0.1439	**0.1682**	0.1662
GBVS'07	0.4580	**0.6099**	0.5980	0.5550	0.6094	**0.6097**	0.5509	**0.6147**	0.6098
PFT'07	0.4273	**0.5526**	0.5460	0.4918	0.5364	**0.5442**	0.4978	0.5490	**0.5575**
IK'98	0.5851	0.6690	**0.6892**	0.6407	0.6708	**0.6800**	0.6374	0.6691	**0.6828**

NSS Method	GAUSS raw	PCA	ZCA	XYZ raw	PCA	ZCA	CAT02LMS raw	PCA	ZCA
CCH'12	0.4916	**0.4942**	0.4875	0.3877	0.4195	**0.4236**	0.3854	0.4194	**0.4273**
QDCT	0.5684	0.5831	**0.5886**	0.4433	0.4981	**0.5382**	0.4477	0.4974	**0.5415**
EPQFT	0.4871	0.4995	**0.5001**	0.3694	0.4183	**0.4738**	0.3729	0.4185	**0.4760**
DCT'11	0.5664	**0.6193**	0.6176	0.4306	**0.5549**	0.5543	0.4332	**0.5553**	0.5545
AC'09	0.1635	0.1629	**0.1663**	0.1225	0.1271	**0.1716**	0.1190	0.1254	**0.1733**
GBVS'07	0.5230	0.6066	**0.6120**	0.3992	**0.5566**	0.5398	0.3989	**0.5589**	0.5357
PFT'07	0.5043	**0.5546**	0.5512	0.3658	**0.5040**	0.4990	0.3696	**0.5051**	0.4984
IK'98	0.6458	0.6735	**0.6777**	0.5245	0.6183	**0.6323**	0.5291	0.6187	**0.6303**

Please refer to Table 3.3 for a color legend

Appendix D
Center Bias Integration Methods

In the main evaluation, see Sect. 4.2.3.2, we present the results that we achieved with a convex combination in Eqs. 4.1 and 4.8. However, we have considered and evaluated alternative integration methods.

To investigate the question how good other combination types are, we tested the minimum, maximum, and product as alternative combinations. To account for the influence of different value distributions within the normalized value range, we also weighted the input of the min and max operation (e.g., $S_P^{min} = \min(w_C S_C, w_B S_B)$). The results of the algorithms using different combination types are shown in Table D.1. The presented results are the results that we achieve with the center bias weight that results in the highest F_1 score.

In Table D.1, we can see that the linear combination is clearly the best choice for LDRC+CB. It is interesting to note that LDRC+CB with the product as combination achieves similar results to RC'10. However, LDRC+CB remains the algorithm that provides the best performance in terms of F_1 score and F_β score whereas RC'10+CB provides the best performance in terms of PHR. Interestingly, LDRC+CB and RC'10+CB achieve a nearly identical $\int$ROC.

© Springer International Publishing Switzerland 2016
B. Schauerte, *Multimodal Computational Attention for Scene Understanding and Robotics*, Cognitive Systems Monographs 30,
DOI 10.1007/978-3-319-33796-8

Table D.1 Salient object detection results that we obtain using different center bias integration types

Method	Combination	F_1	F_β	$\int$ROC	PHR
LDRC+CB	Linear/Convex	<u>0.8034</u>	<u>0.8183</u>	<u>0.9624</u>	0.9240
LDRC+CB	Max	0.7504	0.7561	0.9422	0.8630
LDRC+CB	Min	0.7897	0.8049	0.9535	0.8880
LDRC+CB	Product	0.7883	0.8024	0.9578	0.9130
RC'10+CB	Linear/Convex	0.7973	0.8120	0.9620	0.9340
RC'10+CB	Max	0.7855	0.7993	0.9568	0.9140
RC'10+CB	Min	0.7962	0.8150	0.9603	0.9180
RC'10+CB	Product	0.7974	0.8136	<u>0.9623</u>	<u>0.9460</u>
CB_S	–	0.5793	0.5764	0.8623	0.6980
CB_P	–	0.5604	0.5452	0.8673	0.7120

Please compare to the results in Table 4.1, Sect. 4.2

Creative Commons Image Attribution

Figure 4.16: Bob Bobster (Flickr), "Mailing a letter, 1920s" (CC BY 2.0)
Figure 4.16: Lukas Hron (Flickr), Untitled (CC BY 2.0)
Figure 4.16: Jason Rogers (Flickr), "Day 432/365—It's a small world" (CC BY 2.0)
Figure 4.17: Hans Splinter (Flickr), "near the fire" (CC BY-ND 2.0)
Figure 4.17: Ainis (Flickr), "Happiness" (CC BY-ND 2.0)
Figure 4.17: shira gal (Flickr), "water training" (CC BY 2.0)
Figure 4.17: soldiersmediacenter (Flickr), "iraq" (CC BY 2.0)

References

[AHES09]   Achanta, R., Hemami, S., Estrada, F., Süsstrunk, S.: Frequency-tuned salient region detection. In: Proceedings of the International Conference on Computer Vision and Pattern Recognition (2009)

[Bri95]   Brill, E.: Transformation-based error-driven learning and natural language processing: a case study in part-of-speech tagging. Comp. Ling. 21(4), 543–565 (1995)

[BHCS07]   Becouze, P., Hann, C., Chase, J., Shaw, G.: Measuring facial grimacing for quantifying patient agitation in critical care. In: Computer Methods and Programs in Biomedicine (2007)

[BT09]   Bruce, N., Tsotsos, J.: Saliency, attention, and visual search: an information theoretic approach. J. Vis. 9(3), 1–24 (2009)

[CFK09]   Cerf, M., Frady, E.P., Koch, C.: Faces and text attract gaze independent of the task: experimental data and computer model. J. Vis. 9 (2009)

[CLE]   CLEAR2007, Classification of events, activities and relationships evaluation and workshop. http://www.clear-evaluation.org

[DT05]   Dalal, N., Triggs, B.: Histograms of oriented gradients for human detection. In: Proceedings of the International Conference on Computer Vision and Pattern Recognition (2005)

[Fra79]   Francis, W.N., Kucera, H.: compiled by, A standard corpus of present-day edited american english, for use with digital computers (brown) (1964, 1971, 1979)

[FH04]   Felzenszwalb, P.F., Huttenlocher, D.P.: Efficient graph-based image segmentation. Int. J. Comput. Vis. 59(2), 167–181 (2004)

[GvdBSG01] Geusebroek, J.M., van den Boomgaard, R., Smeulders, A.W.M., Geerts, H.: Color invariance. IEEE Trans. Pattern Anal. Mach. Intell. 23(12), 1338–1350 (2001)

[GCAS+04]   Geoffrey Chase, J., Agogue, F., Starfinger, C., Lam, Z., Shaw, G., Rudge, A., Sirisena, H.: Quantifying agitation in sedated ICU patients using digital imaging. In: Computer Methods and Programs in Biomedicine (2004)

[GMZ08]   Guo, C., Ma, Q., Zhang, L.: Spatio-temporal saliency detection using phase spectrum of quaternion fourier transform. In: Proceedings of the International Conference on Computer Vision and Pattern Recognition (2008)

[HHK12]   Hou, X., Harel, J., Koch, C.: Image signature: highlighting sparse salient regions. IEEE Trans. Pattern Anal. Mach. Intell. 34(1), 194–201 (2012)

[JEDT09]   Judd, T., Ehinger, K., Durand, F., Torralba, A.: Learning to predict where humans look. In: Proceedings of the International Conference on Computer Vision (2009)

[JOvW+05]   Jost, T., Ouerhani, N., von Wartburg, R., Mäuri, R., Häugli, H.: Assessing the contribution of color in visual attention. Comput. Vis. Image Underst. 100, 107–123 (2005)

[KS13]   Koppula, H.S., Saxena, A.: Anticipating human activities using object affordances for reactive robotic response. In: RSS (2013)

[KNd08]   Kootstra, G., Nederveen, A., de Boer, B.: Paying attention to symmetry. In: Proceedings of the British Conference on Computer Vision (2008)

[KSKS12]   Kühn, B., Schauerte, B., Kroschel, K., Stiefelhagen, R.: Multimodal saliency-based
            attention: a lazy robot's approach. In: Proceedings of the 25th International Con-
            ference on Intelligent Robots and Systems (IROS). Vilamoura, Algarve, Portugal:
            IEEE/RSJ, Oct 2012

[KSS13]   Koester, D., Schauerte, B., Stiefelhagen, R.: Accessible section detection for visual
            guidance. In: IEEE/NSF Workshop on Multimodal and Alternative Perception for
            Visually Impaired People (2013)

[LMSR08]   Laptev, I., Marszalek, M., Schmid, C., Rozenfeld, B.: Learning realistic human
            actions from movies. In: Proceedings of the International Conference on Computer
            Vision and Pattern Recognition (2008)

[LSZ+07]   Liu, T., Sun, J., Zheng, N.-N., Tang, X., Shum, H.-Y.: Learning to detect a salient
            object. In: Proceedings of the International Conference on Computer Vision and
            Pattern Recognition (2007)

[MHS10]   Matikainen, P., Hebert, M., Sukthankar, R.: Representing pairwise spatial and tem-
            poral relations for action recognition. In: Proceedings of the European Conference
            on Computer Vision (2010)

[MPK09]   Messing, R., Pal, C., Kautz, H.: Activity recognition using the velocity histories of
            tracked keypoints. In: Proceedings of the International Conference on Computer
            Vision (2009)

[OK04]   Olmos, A., Kingdom, F.A.A.: A biologically inspired algorithm for the recovery
            of shading and reflectance images. Perception **33**, 1463–1473 (2004)

[OSI13]   Onofri, L., Soda, P., Iannello, G.: Multiple subsequence combination in human
            action recognition. IET Computer Vision (2013)

[PLN02]   Parkhurst, D., Law, K., Niebur, E.: Modeling the role of salience in the allocation
            of overt visual attention. Vis. Res. **42**(1), 107–123 (2002)

[PFS12]   Prest, A., Ferrari, V., Schmid, C.: Explicit modeling of human-object interactions
            in realistic videos. IEEE Trans. Pattern Anal. Mach. Intell. **35**, 835–848 (2012)

[PIIK05]   Peters, R., Iyer, A., Itti, L., Koch, C.: Components of bottom-up gaze allocation in
            natural images. Vis. Res. **45**(18), 2397–2416 (2005)

[RR13]   Ren, X., Ramanan, D.: Histograms of sparse codes for object detection. In: Proceed-
            ings of the International Conference on Computer Vision and Pattern Recognition
            (2013)

[RBC06]   Rajashekar, U., Bovik, A.C., Cormack, L.K.: Visual search in noise: revealing the
            influence of structural cues by gaze-contingent classification image analysis. J. Vis.
            6(4), 379–386 (2006)

[RFHS11]   Rybok, L., Friedberger, S., Hanebeck, U.D., Stiefelhagen, R.: The KIT robo-kitchen
            data set for the evaluation of view-based activity recognition systems. In: Hu-
            manoids (2011)

[RSAHS14]   Rybok, L., Schauerte, B., Al-Halah, Z., Stiefelhagen, R.: Important stuff, every-
            where! Activity recognition with salient proto-objects as context. In: Proceedings
            of the 14th IEEE Winter Conference on Applications of Computer Vision (WACV),
            Steamboat Springs, CO, USA, March 2014

[RDM+13]   Riche, N., Duvinage, M., Mancas, M., Gosselin, B., Dutoit, T.: Saliency and human
            fixations: state-of-the-art and study of comparison metrics. In: Proceedings of the
            International Conference on Computer Vision (2013)

[SG31]   Smith, T., Guild, J.: The C.I.E. colorimetric standards and their use. Trans. Opt.
            Soc. **33**(3), 73 (1931)

[SF10a]   Schauerte, B., Fink, G.A.: Focusing computational visual attention in multi-modal
            human-robot interaction. In: Proceedings of the 12th International Conference on
            Multimodal Interfaces and 7th Workshop on Machine Learning for Multimodal
            Interaction (ICMI-MLMI). ACM, Beijing, China, Nov 2010

[SF10b]   Schauerte, B., Fink, G.A.: Web-based learning of naturalized color models for
            human-machine interaction. In: Proceedings of the 12th International Conference
            on Digital Image Computing: Techniques and Applications (DICTA). IEEE, Syd-
            ney, Australia, Dec 2010

[SRF10]   Schauerte, B., Richarz, J., Fink, G.A.: Saliency-based identification and recognition of pointed-at objects. In: Proceedings of the 23rd International Conference on Intelligent Robots and Systems (IROS). IEEE/RSJ, Taipei, Taiwan, Oct 2010

[SPSS12]   Sung, J., Ponce, C., Selman, B., Saxena, A.: Unstructured human activity detection from RGBD images. In: Proceedings of the International Conference on Robotics and Automation (2012)

[TMZ+07]   Temko, A., Malkin, R., Zieger, C., Macho, D., Nadeu, C., Omologo, M.: Clear evaluation of acoustic event detection and classification systems. In: Stiefelhagen, R., Garofolo, J. (eds.) ser. Lecture Notes in Computer Science, vol. 4122, pp. 311–322. Springer, Berlin (2007)

[VW87]   Vallacher, R.R., Wegner, D.M.: What do people think they're doing? action identification and human behavior. Psychol. Rev. **94**(1), 3–15 (1987)

[vdWSV07]   van de Weijer, J., Schmid, C., Verbeek, J.J.: Learning color names from real-world images. In: Proceedings of the International Conference on Computer Vision and Pattern Recognition (2007)

[WCW11]   Wang, J., Chen, Z., Wu, Y.: Action recognition with multiscale spatio-temporal contexts. In: Proceedings of the International Conference on Computer Vision and Pattern Recognition (2011)

[YL13]   Yi, Y., Lin, Y.: Human action recognition with salient trajectories. Signal Processing (2013)

Printed in the United States
By Bookmasters